Assessment of the USCENTCOM Medical Distribution Structure

William Welser IV, Keenan D. Yoho, Marc Robbins, Eric Peltz,

Ben D. Van Roo, Adam C. Resnick, Ronald E. Harper

Prepared for the United States Army

Approved for public release; distribution unlimited

RAND ARROYO CENTER

The research described in this report was sponsored by the United States Army under Contract No. W74V8H-06-C-0001.

Library of Congress Cataloging-in-Publication Data

Assessment of the USCENTCOM medical distribution structure / William Welser, IV [et al.].
  p. cm.
 Includes bibliographical references.
 ISBN 978-0-8330-4923-0 (pbk. : alk. paper)
 1. United States--Armed Forces--Medical supplies. 2. United States--Armed Forces--Supplies and stores. 3. United States. Central Command. I. Welser, William.

UH443.A884 2010
355.8'8--dc222
                                                          2010007761

The RAND Corporation is a nonprofit research organization providing objective analysis and effective solutions that address the challenges facing the public and private sectors around the world. RAND's publications do not necessarily reflect the opinions of its research clients and sponsors.

RAND® is a registered trademark.

Published 2010 by the RAND Corporation
1776 Main Street, P.O. Box 2138, Santa Monica, CA 90407-2138
1200 South Hayes Street, Arlington, VA 22202-5050
4570 Fifth Avenue, Suite 600, Pittsburgh, PA 15213-2665
RAND URL: http://www.rand.org/
To order RAND documents or to obtain additional information, contact
Distribution Services: Telephone: (310) 451-7002;
Fax: (310) 451-6915; Email: order@rand.org

Preface

This study examined whether there is a less costly medical distribution structure for U.S. Central Command (USCENTCOM) that would maintain or improve performance. The assessment considered five options, evaluating the likely performance and cost implications as well as any effects on related nondistribution activities. The first option is the status quo of supporting USCENTCOM from a combination of U.S. Army Medical Materiel Center, Southwest Asia (USAMMC-SWA), located in Qatar, and U.S. Army Medical Materiel Center, Europe (USAMMCE), located in Germany. The second option is to support USCENTCOM directly from the continental United States (CONUS) prime vendor support with shipments sent through the Defense Distribution Depot Susquehanna, Pennsylvania (DDSP) containerization and consolidation point (CCP). The third option is to stock medical materiel at the nonmedical distribution depot in Kuwait instead of separately in Qatar. The fourth option would be to support USCENTCOM solely from USAMMCE. The fifth is to increase the breadth of stocks at USAMMC-SWA so that it could provide almost all direct support to USCENTCOM customers.

This research was sponsored by the Commanding General of the U.S. Army Medical Research Materiel Command. It should be of broad interest to Department of Defense supply chain managers, logisticians, and medical personnel. This research has been conducted within RAND Arroyo Center's Military Logistics Program. RAND Arroyo Center, part of the RAND Corporation, is the Army's federally funded research and development center for policy studies and analyses.

The Project Unique Identification Code (PUIC) for the project that produced this document is DASGP09198.

Questions and comments regarding this research are welcome and should be directed to the director of the Military Logistics Program, Eric Peltz, at Eric_Peltz@rand.org, or to Bill Welser, at bill_welser@rand.org.

For more information on RAND Arroyo Center, contact the Director of Operations (telephone 310-393-0411, extension 6419; fax 310-451-6952, email Marcy_Agmon@rand.org), or visit Arroyo's web site at http://www.rand.org/ard.

Contents

Figures

Tables

Summary

In July 2008, the Director of Logistics of the Joint Staff (JSJ4) and the U.S. Army Deputy Chief of Staff, G-4 (Army G-4) visited the U.S. Central Command (USCENTCOM) area of responsibility (AOR) to review logistics operations. One of their questions was whether efficiencies could be gained by combining medical and nonmedical warehouse distribution with stocks consolidated at one location. RAND Arroyo Center and U.S. Army Medical Research and Materiel Command (USAM-RMC) expanded this question and explored whether there might be less costly medical distribution structures for USCENTCOM that would maintain the quality of health care delivery. In this report we describe the current distribution structure for medical (Class VIII) materiel for USCENTCOM customers, a set of alternatives, and the likely performance, cost, and other effects of changing the current system to that of one of the alternatives.

Background: The Current Distribution Structure and Its Origins

Class VIII materiel is supplied to USCENTCOM AOR customers from two distribution centers: U.S. Army Medical Materiel Center, Southwest Asia (USAMMC-SWA), located at Camp As Sayliyah, Qatar; and U.S. Army Medical Materiel Center, Europe (USAMMCE), located in Pirmasens, Germany. Approximately 60 percent of the medical materiel sent to USCENTCOM AOR customers comes from USAMMC-SWA, which stocks 3,000 lines of the fastest-moving items and is replenished by USAMMCE. The other 40 percent of the requisitions that cannot be filled by USAMMC-SWA are passed back to and filled directly by USAMMCE, which carries approximately 13,000 lines of stock and is replenished by commercial prime vendors.

Non-Class VIII materiel for USCENTCOM customers is shipped primarily from the Defense Distribution Depot Kuwait, Southwest Asia (DDKS), from continental U.S. (CONUS) distribution centers—most often Defense Distribution Depot Susquehanna, PA (DDSP)—and directly from vendors for certain classes of items such as food. Managed by the Defense Logistics Agency (DLA), DDKS is a contractor-owned and -operated distribution center that stores and distributes supply Classes II (such as textiles, uniforms, tents), IIIp (packaged petroleum products), IV (barrier and construction materials), and IX (repair parts). Figure S.1 shows the locations for DDKS, USAMMCE, and USAMMC-SWA.

Initially, USCENTCOM nonmedical theater-level sustainment stocks were stored in Army general support (GS) supply support activities (SSAs), which were stood up in early 2003. When DDKS became active in 2004, the Army phased out the inventory levels in the Class II, IIIp, and IV and Class IX common GS SSAs so that they would not be replenished. However, a GS SSA remained in place to accept and process serviceable returns. In 2007, shipments from the GS SSA and DDKS were combined on pallets to improve distribution performance and improve transportation efficiency through larger pallets built more quickly. In 2008, process and information system changes were made so that DDKS could take over the mission of receiving and processing serviceable returns for increased warehousing and distribution efficiency. The new question was whether further efficiencies could be gained by also consolidating distribution of non-Class VIII and Class VIII medical supplies.

Figure S.1
Locations of USAMMCE, USAMMC-SWA, and DDKS

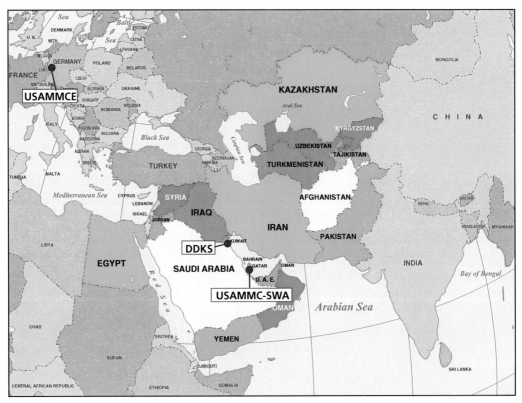

Study Methodology and Evaluation of Options

The initial options were to consider supporting the USCENTCOM AOR with medical materiel shipped via prime vendor support from CONUS through DDSP, stocking medical supplies at DDKS and closing the distribution center at USAMMC-SWA, and the status quo structure (see Table S.1). Two additional options were considered based upon preliminary data analysis and as a result of interviews with medical logisticians: providing all direct support of medical materiel to USCENTCOM from USAMMCE, and replicating more of the USAMMCE inventory at USAMMC-SWA so that it could directly provide most items to customers.

The first criterion that each option had to satisfy was that of performance: Does the option maintain or improve performance with regard to how long it takes to fill orders? The medical supply chain is focused on clinical outcomes, and timely response to needs is considered critical, with current performance considered acceptable. Thus, the medical community expected that the performance for any new distribution option would be equal to or better than current performance.[1] Second, is the option less costly than the status quo? If an option meets these two criteria of performance and cost, then it is considered a possible distribution option for medical materiel to the USCENTCOM AOR (see Figure S.2).

The CONUS Option

The best representation of the time associated with this option is the time for direct vendor delivery (DVD) shipments from CONUS, since there is no CONUS stockage of medical supplies—only prime vendor support. We compared average end-to-end distribution times for Class VIII and DVD shipments of Class IX materiel through the DDSP CCP to customers in USCENTCOM (see Figure S.3).[2] Beginning at a vendor

Table S.1
Medical Distribution Options Considered

Option	Performance	Cost	Other
Status quo			
Prime vendor from CONUS through DDSP CCP			
Stock medical supplies at DDKS			
Consolidate at USAMMCE (no USAMMC-SWA)			
Replicate USAMMCE stocks at USAMMC-SWA			

---

[1] Performance is measured in terms of distribution time.

[2] The DVD model is the closest analogue to the current medical model that relies upon prime vendor support.

Figure S.2
Decision Framework for Evaluating Distribution Options

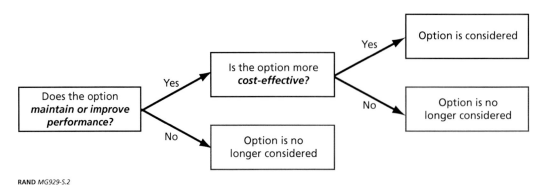

Figure S.3
Average Distribution Time Performance from CONUS

MRO to destination arrival for USAMMC-SWA and Class IX DVDs from CONUS to CENTCOM Customers

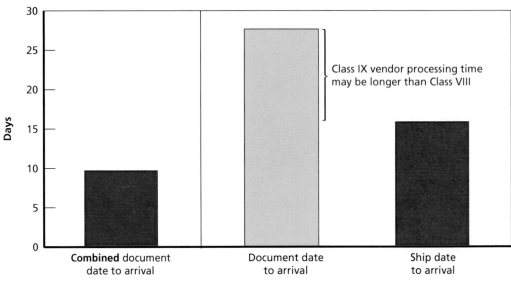

location, DVD Class IX materiel is shipped to the CCP and then forwarded on to the final aerial port of debarkation (APOD). The average time from the moment a materiel release order (MRO) goes to the vendor for a DVD item until the materiel arrives at the customer's APOD when originating from CONUS is 28 days, whereas Class VIII average performance is 10 days total time. If we take out the vendor processing time to account for the possibility that medical prime vendors have better processes than the non-Class VIII DVD suppliers, the average time from shipment to arrival for DVD shipments is still longer than the total Class VIII average distribution time. Because the performance associated with the CONUS option is significantly worse than that associated with the current medical materiel distribution structure, we did not consider this option further.

The DDKS Option

The data in Figure S.4 show that USAMMC-SWA has a performance advantage over DDKS. Times from MRO to arrival at the APOD from DDKS have averaged about 6.5 days in fiscal year (FY) 2009, whereas USAMMC-SWA shipments averaged 4 days. The advantage for USAMMC-SWA lies primarily in the MRO-to-pick segment, with some advantage also in the transportation segment.

To compare costs between the two locations, we estimated how much it would cost to conduct the USAMMC-SWA distribution center mission at DDKS, the actual costs for performing this mission at USAMMC-SWA, and how transportation costs would change based upon the actual airlift rates from the two locations.

Figure S.4
Average Distribution Segment Times: DDKS and USAMMC-SWA

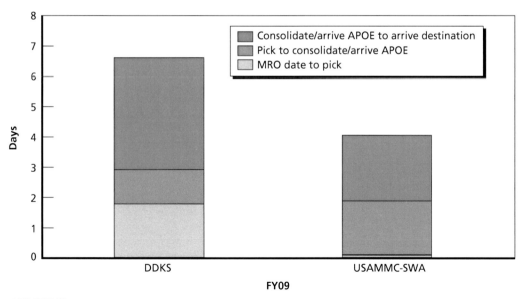

RAND *MG929-S.4*

We combined the DDKS cost and volume data to develop cost factors (or rates) to estimate the DDKS operating costs of performing the USAMMC-SWA mission. We also estimated what construction would cost at DDKS if it were determined that additional space would be needed for the medical mission, and we provided cost estimates with and without the construction costs due to uncertainty with regard to this requirement.

To calculate the transportation difference, we decomposed the weight moved by USAMMC-SWA country (Iraq and Afghanistan) and by month, and then we applied the appropriate airlift rates by destination country from the two depots. Figure S.5 shows the cost estimates.

A cost-sensitivity analysis was conducted by increasing the number of pounds of Class VIII materiel shipped to Afghanistan and decreasing the Class VIII pounds shipped to Iraq, in accordance with FY2009 trends and Department of Defense (DoD) planning (see Figure S.6). For each set of conditions there are two bars, blue and red. The lower, blue bars correspond to the operating cost estimates that assume no con-

Figure S.5
Cost Estimates for USAMMC-SWA and DDKS

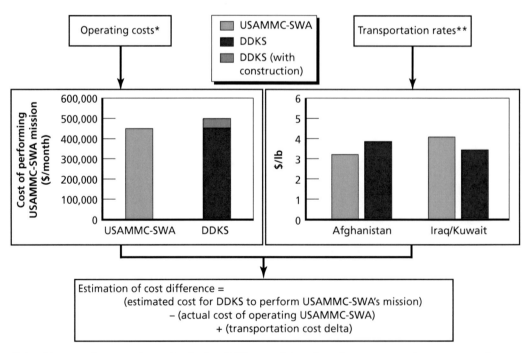

*Monthly operating costs determined using FY08 annual data, based upon transaction costs at each location.

**Transportation rates using October–December 2008 data; rates are not at steady state as USAMMC-SWA ratio of tender to total continues to increase.

RAND MG929-S.5

Figure S.6
Cost Sensitivity Analysis for USAMMC-SWA and DDKS

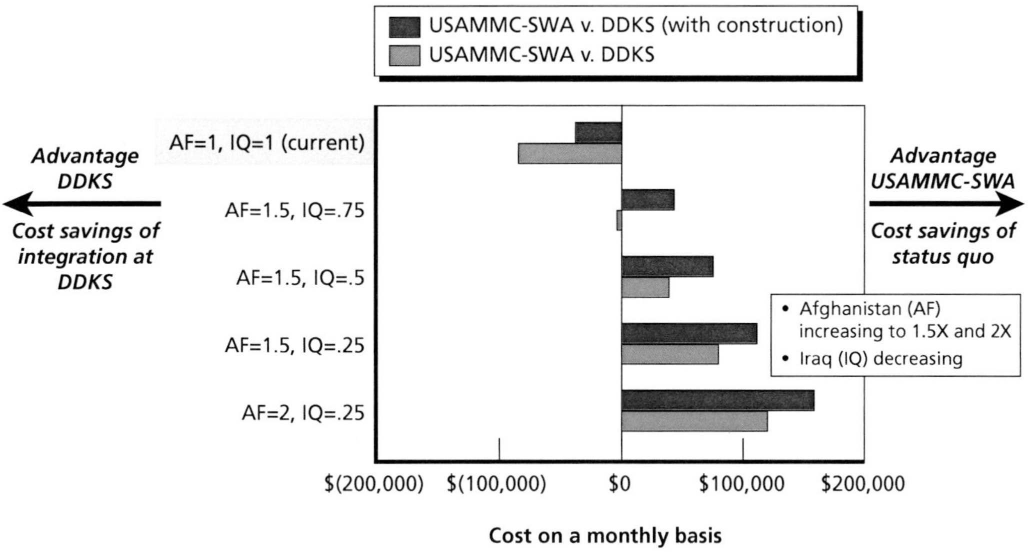

RAND *MG929-S.6*

struction costs are necessary at DDKS for Class VIII mission absorption. The top, red bars represent the cases where the cost of new construction is included in the estimate of DDKS operating rates. A bar displaying value to the right of the center vertical axis indicates that there is a cost advantage to continuing to supply USCENTCOM Class VIII materiel from USAMMC-SWA. Conversely, a bar pointing to the left indicates that performing the USCENTCOM Class VIII mission from DDKS would generate a cost savings over continuing the mission from USAMMC-SWA.

Under "current" conditions, DDKS is estimated to have a slight cost advantage over USAMMC-SWA.[3] However, as troop levels in Afghanistan increase and Iraq levels decrease, the cost difference shifts to favor USAMMC-SWA. When the Afghanistan weight is doubled and Iraq is at one-quarter, we estimate that the cost advantage for USAMMC-SWA would reach up to $160,000 per month.

The Option of Consolidating Operations at USAMMCE

While distribution times from USAMMCE appear to be longer than those from USAMMC-SWA, these differences are driven by their respective roles in the USCENT-

---

[3] "Current" conditions are defined by taking the average of the Class VIII weights shipped during October–December 2008.

COM supply chain and not by process performance differences (see Figure S.7).[4] We found that if USAMMCE had USAMMC-SWA's direct customer support mission for USCENTCOM, the times would most likely be similar.

The two main factors driving the performance differences between the two locations are that USAMMCE is on a five-day work week instead of seven, and the fact that USAMMCE is the second source of supply. Being the second source creates requisition pass-back delays, exacerbated by batching, and lower volume, which leads to longer time to collect materiel for consolidated shipments. There are also some backorders miscoded as immediate issue shipments in the USAMMCE data (USAMMC-SWA has no backorders, because they are all passed to USAMMCE).

When USAMMCE is the primary source of support for customers, performance looks similar to that of USAMMC-SWA for CENTCOM customers. The column on the far right of the chart shows FY2008 performance for USAMMCE in support of

Figure S.7
Average Distribution Segment Times: USAMMCE and USAMMC-SWA

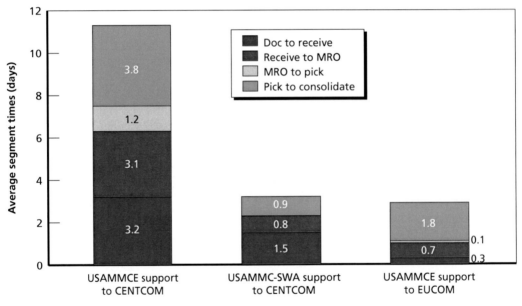

RAND *MG929-S.7*

---

[4] The first segment (*doc to receive*) covers the time from the initial request for materiel, or document date (*doc*), to the time that the request is received electronically at the depot (*receive*). The next segment (*receive to MRO*) is the time that it takes for the *received* request to be prioritized and printed as an MRO for the depot to *issue*. The third segment (*MRO to pick*) is the time that it takes for the warehouse staff to identify and collect the materiel requested. The fourth segment is the amount of time necessary to consolidate all materiel to be shipped to a particular customer(s) or location(s).

major U.S. European Command (USEUCOM) customers. Overall times excluding transportation are roughly the same as for USAMMC-SWA for its USCENTCOM customers.[5]

Figure S.8 depicts the transportation structure for USAMMCE and USAMMC-SWA as of FY2009. USAMMCE uses a Class VIII commercial tender to move a high percentage (greater than 98 percent) of the total weight it ships to customers in USCENTCOM. USAMMC-SWA uses the Class VIII tender for roughly 40 percent of its shipments, in terms of weight. Note that the Class VIII tender shipments for both distribution centers are shipped through the same carrier hub for final shipment to the destination airfield, and replenishments from USAMMCE to USAMMC-SWA are shipped through this same hub as well. The transportation structure drives most of the difference in the distribution costs between the two locations.

On the left-hand side of Figure S.9, we show USAMMC-SWA's operating costs per month. In the same graph, we show the estimated monthly operating cost increase at USAMMCE were it to perform USAMMC-SWA's mission. The total weight shipped would not change, because replenishment shipments would merely shift to customer issues. However, there would be an increase in transactions due to smaller

Figure S.8
FY2009 Transportation Structure for USAMMCE and USAMMC-SWA

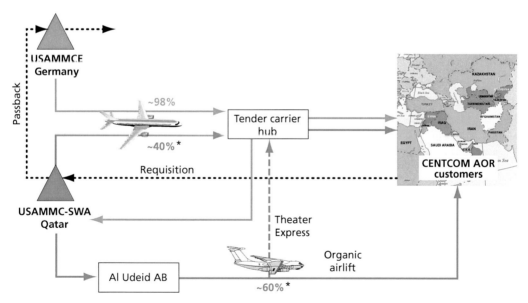

*USAMMC-SWA ratio of tender to total rose in January–February 2009 to ~55%.
RAND MG929-S.8

---

[5]   We have no means of measuring transportation times from USAMMCE to its USEUCOM customers, so we exclude showing the transportation segment in all three cases in Figure S.7.

Figure S.9
USAMMCE and USAMMC-SWA Operating and Transportation Costs

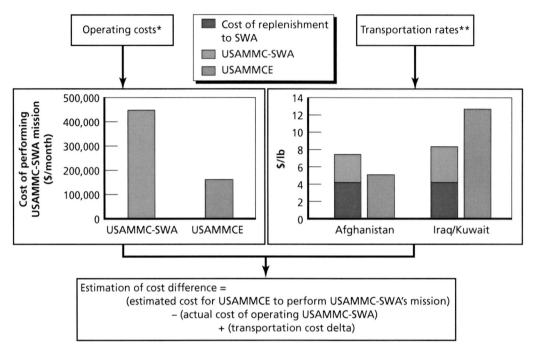

*Monthly operating costs determined using FY08 annual data and transaction costs at each location with replenishments to SWA replaced by USAMMCE issues.

**Transportation rates using October–December 2008 data; rates are not at steady state as USAMMC-SWA ratio of tender to total continues to increase.

RAND MG929-S.9

quantities per transaction. Additionally and more importantly, USAMMCE would need to move to a seven-day work week to achieve USAMMC-SWA–like performance. We take these changes into account.

On the right-hand side of Figure S.9 is a graph comparing the aggregate transportation rates from the two distribution depots. The dark blue lower portions of the left columns indicate the cost of replenishments to USAMMC-SWA, and the light blue upper portions show the cost of actually going from USAMMC-SWA to the customer. For airlift to Afghanistan, it is less expensive to ship from USAMMCE. However, under the current structure and the commercial carrier selections for Iraq, it is less expensive to replenish USAMMC-SWA from USAMMCE and then ship to the customer than it is to simply ship directly from USAMMCE to the customer.

Figure S.10 shows the effect of increasing shipments to Afghanistan and decreasing them to Iraq. With an increase in troop levels in Afghanistan to 1.5 times early

Figure S.10
Cost Sensitivity Analysis for USAMMCE and USAMMC-SWA

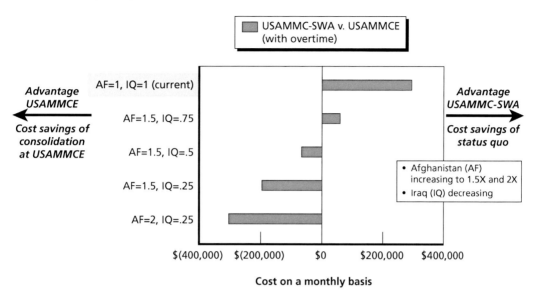

NOTE: Cost of replenishment is included for SWA for all cases.

RAND *MG929-S.10*

FY2009 levels and Iraq levels decreasing in 25-percent steps from the early FY2009 levels, the cost comparison shifts from favoring the status quo to favoring consolidation at USAMMCE. If the troop strength in Afghanistan were to double from early FY2009 levels and troop levels in Iraq were to fall to one-quarter of early FY2009 levels, there would be an estimated savings of roughly $300,000 per month associated with consolidating distribution support for USCENTCOM at USAMMCE.

As shown in Table S.2, there are several other capabilities at USAMMC-SWA other than materiel warehousing and distribution, namely: medical equipment maintenance/repair, forward repair activity mission (FRA-M) support, patient movement item (PMI) cell support, optical fabrication, and customer technical support. However, medical maintenance actions conducted at USAMMC-SWA could probably be absorbed into the existing USAMMCE maintenance operations. Additionally, centralizing repair parts inventory at one location could reduce the overall cost of this inventory. Centralizing repair technicians could also facilitate cross-training among the workforce and provide more time on equipment for repair experience. We did not find any data that would indicate performance degradation or an increase in costs if the FRA-M teams, PMI cell support, optical fabrication, or customer support were not located at USAMMC-SWA.

Table S.2
Capabilities at USAMMC-SWA

SWA Capability	Option	Implications		
		Performance	Cost	Intangibles
Medical equipment maintenance and repair	Move to USAMMCE	No known impact	May reduce cost of repair part inventory	May increase cross-training; will have access to ISO 9000 facilities
FRA-M mission support	Move to USAMMCE or Balad	No known impact	No known impact	The FRA-M team only needs a bed-down location
Patient movement item (PMI) cell support	Move to point of sortie origin or destination (i.e., Ramstein)	No known impact	No known impact	
Optical fabrication	Move to USAMMCE	No known impact	No known impact	
Customer and contingency operations support	Move to USAMMCE	No known impact	No known impact	May not have support that is fully "attuned" to theater environment

The Option of Replicating USAMMCE Inventory at USAMMC-SWA

Just as consolidating support at USAMMCE, replication of USAMMCE capabilities at USAMMC-SWA so that most customer shipments would come from there would improve performance by eliminating distribution network fragmentation, but there would be some cost penalty. Inventory investments would have to be made at USAMMC-SWA. Currently, USAMMCE stocks approximately 13,000 lines of materiel, while USAMMC-SWA stocks approximately 3,000 of the fastest-moving lines. Based upon a rough estimate, an 85 percent customer demand fill rate target would require approximately 5,600 additional lines (for a total of 8,600 lines to be stocked at USAMMC-SWA) at a total "buildup" cost that would likely be less than $1 million.[6] One potential complication is that if additional inventory were added to USAMMC-SWA and a customer service fill rate target of 85 percent were achieved, 15 percent of orders would still have to be satisfied by USAMMCE. This low volume of materiel might be a problem for the Class VIII tender usage by USAMMCE—as it might not be enough for the service or to get the prices that are in effect at this time, and alternatives such as the general USTRANSCOM World Wide Express contract might have to be explored.

---

[6] Although there may not be adequate space to accommodate the additional stock levels at USAMMC-SWA at its current Camp As Sayliyah location, there is a request in to add an additional 30,000 square feet of space when the operations are moved to Al Udeid Air Base by the fourth quarter of 2012.

In addition to inventory, there would be a need for personnel to manage the medical air bridge supplying materiel coming out of CONUS, assuming direct replenishment as opposed to replenishment from USAMMCE stocks, as well as personnel to manage the new item requests (NIRs), which number in the hundreds per month at USAMMCE. An alternative to locating the vendor support and NIR processing forward could be to establish a capability within CONUS to remotely perform these activities.

The Value of Consolidation at a Single Location

Supporting CENTCOM customers out of one location would likely yield better performance by eliminating the delays associated with split-sourcing. The potential benefits of doing so could be a 20 percent improvement in average end-to-end time, as shown in Figure S.11. If USAMMCE were the sole source, it would be necessary to move to a seven-day per week schedule to support ongoing war operations. Alternatively, USAMMC-SWA could be the predominant source. As noted, in this case, USAMMCE's distribution strategy for direct support to USCENTCOM customers would most likely have to change.

Figure S.11
Estimated Average Distribution Time Associated with Single Location

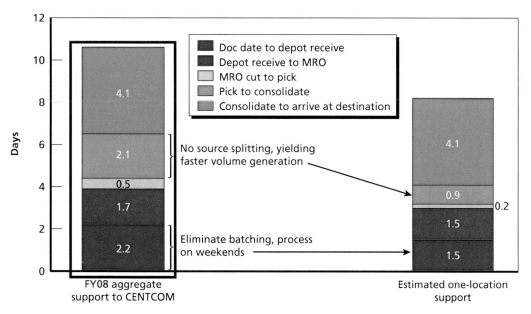

RAND *MG929-S.11*

Acceptable Medical Distribution Options

There are three options that would preserve or improve performance while maintaining or lowering costs (see Table S.3). Consolidation at one location would yield 20 percent better performance, and, if at USAMMCE, consolidation would likely provide for a relatively modest reduction in total costs, anywhere from $1 million to $3.5 million per year. Such consolidation could potentially further reduce costs and improve performance through renegotiation of the Class VIII tender contracts to provide all materiel distribution out of one airfield. Consolidation at USAMMC-SWA would improve performance, perhaps even more, but it would be more costly than consolidation at USAMMCE, as inventory would increase, economies of scale would not be leveraged, and transportation rates to Afghanistan would be higher from this location. The status quo would not change cost or performance.

Since distribution performance would be worse and costs would not be lower than USAMMC-SWA, the DDKS option does not meet the criteria for an option to be considered. Support from CONUS through a CCP would result in unacceptable performance and is therefore not an option that would meet the criteria for consideration.

Table S.3
Assessment of Distribution Options

Option	Performance	Cost	Other Factors
Status quo	—	—	—
Consolidate at USAMMCE	Slightly better performance than status quo with elimination of pass-back delays and consolidation	Better cost efficiency	—
Replicate at USAMMC-SWA	Better performance than status quo	Potentially higher cost	Would need to establish and manage prime vendor support and new item request management[a]
DDKS	Worse performance to Afghanistan and Iraq	Likely similar cost; some risk of higher cost	Transition would create need for medical logistics, specific assets, and medical logistics information system
CONUS support: DDSP	Overall worse performance	—	Transition would create need for medical logistics, specific assets, and medical logistics system

NOTE: Shaded areas do not meet acceptability criteria.

[a] Establishing a CONUS capability to provide prime vendor and NIR support for deployed units might mitigate this personnel and management requirement.

Acknowledgments

There were many organizations and individuals who made this study possible through their hard work, willingness to track down critical data and information, and dedication to its independent and objective outcome. We would like to recognize them here and thank them for their assistance and commitment to ensuring the work was thorough and timely.

COL Edmund Haraguchi helped shape the initial questions that were the basis of the study, provided ongoing enthusiastic support, and helped to ensure that there was visibility within the medical logistics community as to the ongoing work and its findings. COL Michael P. Ryan, Deputy Assistant Chief of Staff for Logistics, U.S. Army Medical Command, provided important feedback and suggestions in shaping the study from the first meeting at Fort Detrick, and his suggestion to meet with medical logisticians from the field during a conference in Qatar was important in developing an understanding of performance needs.

We would like to thank Rear Admiral Mark F. Heinrich, then Deputy Commander of United States Central Command Deployment Distribution Operations Center (CDDOC), for his time and access to his command to answer questions and contribute to the work. We would also like to thank the commanders of USAMMCE, COL Mitchell Brew, and USAMMC-SWA, LTC Sam Haddad, for their generosity during our site visits, and for their support and openness during the course of the study. Both COL Brew and LTC Haddad improved the quality of the study by sharing the time of their respective teams as well as their own professional insights and subject matter expertise. We would also like to thank COL Clayton T. Newton, commander of DDKS, and his deputy Mr. Jerry Brown. Additionally, we would like to thank Mr. William Stenhouse and his team at Agility PWC who execute the operations at DDKS.

Under Mr. David Williams' direction, this project moved forward unobstructed. There was no stone too large to overturn, no avenue of inquiry too treacherous to explore, no institutional thicket of procedures or regulations too dense for him to negotiate, and no timeline too severe for him to spur on those around him to ensure an objective and thorough investigation of the data and facts. We thank him for his good humor, dedication, and fire of spirit. LTC Song Gotiangco at the USAMRMC was very supportive throughout the study, assisting with protocol and keeping the RAND Arroyo Center team closely connected to Mr. Williams.

The entire research team benefited from LTC David Gibson's willingness to immerse himself immediately in not only the intellectual but also administrative aspects of the study. His incisive questions, refreshing insights, enthusiasm for the research, and gracious demeanor were very important factors throughout.

COL (ret.) Jon Kissane is recognized and respected throughout the medical community as an intellectual leader and architect of many of the organizational structures and process initiatives that currently support U.S. combat medicine. The Arroyo research team benefited from his knowledge, historical context, and in-depth discussion throughout the study's duration. His nuanced understanding of the historical and contemporary context of the medical logistics and distribution structure helped the research team understand some of the broader implications of changes to any part of this functioning and performing system.

Mr. Michael O'Connor of the Defense Supply Center Philadelphia was a critical "Center of Gravity" during the early field work, particularly in Kuwait where he helped the Arroyo team interface with representatives from Defense Logistics Agency, the U.S. Navy, and the U.S. Army Central Command Surgeon General's Office.

Two individuals whose assistance was critical throughout were LTC William M. Stubbs of USAMMCE and MAJ Jennifer Allouche of USAMMC-SWA. The Arroyo team's unquenchable thirst for data and details on distribution operations was met with gracious support, fast responses, and thorough answers thanks to these two individuals. Both LTC Stubbs and MAJ Allouche have made considerable contributions to their respective organizations, and it was their willingness to share the ongoing changes in business processes and policies that allowed the research team to get the most up-to-date picture of USCENTCOM Class VIII distribution operations.

COL Michael S. McDonald, commander of the 6th Medical Logistics Management Center, provided access to his team of professionals who were instrumental in providing the team with access to the Theater Army Medical Management Information System. Specifically, MAJ Cynthia Hammer and Mr. John Sprowls provided distribution data and helped the team with its interpretation of data fields and medical business processes. Without the contribution of these two individuals, none of the distribution process analysis would have been possible.

CW4 Karen Droessler was instrumental in directing the team to the sources of medical maintenance data, and both she and CW2 Kenneth Bynums assisted in its interpretation.

There were many others whose contributions to our research effort were timely and valuable. These included MAJ Brandon Pretlow and MAJ Willie Davis at USAMMC-SWA as well as LtCol Aaron Gittner at the Tanker Airlift Control Center.

We would also like to express appreciation to the kind medical professionals who met with us during the October 2008 USCENTCOM Medical Logistics conference at Camp As Sayliyah: CPT Watson, CPT Reyes and 1SG Alvarado, MAJ Michael, 1LT Ware, HM Marquis and Lieutenant Flores, MAJ Haug, CW Furr, and CPT Barr.

We also owe gratitude to the excellent peer reviewers of this document, Mr. Peter Lukszys of the University of Wisconsin-Madison and Dr. Ronald McGarvey of RAND.

Several members of RAND Arroyo Center were important in bringing this study to a timely, high-quality conclusion. The expertise and guidance shared by Eric Peltz, the director of RAND Arroyo Center's Military Logistics Program, were extremely important; his contributions and direction, both thoughtful and deliberate, were key factors in delivering this document in such a manner that it was both thorough and timely. Rick Eden provided constructive feedback and criticism throughout the study which helped to sharpen and clarify much of the work presented during briefings to senior leaders. Kristin Leuschner and Pamela Thompson aided the team with document editing, while Todd Duft, Nikki Shacklett, and Benson Wong deserve the utmost credit for playing the role of "document magicians" in the production editing and final publication process. Holly Johnson expertly guided the document through the final stages of preparation for publication in the smoothest fashion possible. Finally, Patrice Lester deserves the utmost credit for keeping our team traveling, writing, meeting, "VTCing," laughing, and breathing throughout the short project timeline of six months.

Acronyms

AB	Air Base
AIED	A transportation account code (TAC)
AILD	A transportation account code (TAC)
AMC	Air Mobility Command
AMD	Air Mobility Division
AOR	Area of Responsibility
APOD	Aerial Port of Debarkation
APOE	Aerial Port of Embarkation
APS	Army Prepositioned Stocks
C2	Command and Control
CAOC	Combined Air Operations Center
CCP	Containerization and Consolidation Point
CDDOC	CENTCOM Deployment Distribution Operations Center
Class II	Clothing
Class IIIp	Petroleum
Class IV	Fortification and Barrier Materials
Class VIII	Medical Supplies
Class IX	Repair Parts
CONUS	Continental United States
DC	Distribution Center
DCSLOG	Deputy Chief of Staff for Logistics
DDKS	Defense Distribution Depot Kuwait, Southwest Asia
DDSP	Defense Distribution Depot Susquehanna, Pennsylvania
DFAS	Defense Finance and Accounting Service
DLA	Defense Logistics Agency

DoD	Department of Defense
DRA	Defense Reporting Activity
DSCP	Defense Supply Center Philadelphia
DVD	Direct Vendor Delivery
EAMS	Expeditionary Air Mobility Squadron
EDI	Electronic Data Interchange
EMF	Expeditionary Medical Facility
FDD	Forward Distribution Depot
FRA-M	Forward Repair Activity Mission
GATES	Global Air Transportation Execution System
GBL	Government Bill of Lading
GS	General Support
ILAP	Integrated Logistics Analysis Program
JSJ4	Director of Logistics of the Joint Staff
LTG	Lieutenant General
MilAir	Military Airlift
MILSTRIP	Military Standard Requisitioning and Issue Procedures
MLMC	Medical Logistics Management Center
MRO	Materiel Release Order
NAC	National Air Cargo
NIR	New Item Request
OEF	Operation Enduring Freedom
OIF	Operation Iraqi Freedom
ORF	Operational Readiness Float
OTSG	Office of the Surgeon General
PEO-EIS	Program Executive Office–Enterprise Information Systems
PMI	Patient Movement Items
POTUS	President of the United States
PV	Prime Vendor
RFID	Radio Frequency Identification Data
SDP	Strategic Distribution Platform
SSA	Supply Support Activity
TAC	Transportation Account Code
TACC	Tanker Airlift Control Center

TAMMIS	Theater Army Medical Management Information System
TCN	Transportation Control Number
TEWLS	TAMMIS Enterprise Wide Logistics System
TWCF	Transportation Working Capital Fund
USAF	United States Air Force
USAFRICOM	United States Africa Command
USAMMA	United States Army Medical Materiel Agency
USAMMCE	United States Army Medical Materiel Center, Europe
USAMMC-SWA	United States Army Medical Materiel Center, Southwest Asia
USAMRMC	United States Army Medical Research and Materiel Command
USCENTCOM	United States Central Command
USEUCOM	United States European Command
USTRANSCOM	United States Transportation Command

CHAPTER ONE

Introduction

Medical logistics is a distinct and separate function and organization from other logistics operations within the U.S. Army. Past studies have typically concluded that Class VIII (medical) supply is sufficiently unique and different from other supply classes to call for separate handling, distribution, and management,[1] yet this separation is periodically questioned and re-evaluated. Demand for medical materiel is often urgent; further, there are legal mandates that govern the storage and control of many medical supplies (such as narcotics). In addition, some of the products are vulnerable to temperature changes, exposure to the elements, or degradation over time.

Class VIII materiel is supplied to customers in the U.S. Central Command (USCENTCOM) area of responsibility (AOR) from two distribution centers: U.S. Army Medical Materiel Center, Southwest Asia (USAMMC-SWA), located at Camp As Sayliyah, Qatar, and U.S. Army Medical Materiel Center, Europe (USAMMCE), located in Pirmasens, Germany. Approximately 60 percent of the medical materiel sent to USCENTCOM AOR customers comes from USAMMC-SWA. The other 40 percent of requisitions that cannot be filled by USAMMC-SWA are passed back to and filled directly by USAMMCE, which is replenished by commercial prime vendors.

In July 2008, the Director of Logistics of the Joint Staff (JSJ4) and the U.S. Army Deputy Chief of Staff, G-4 (Army G-4) visited the USCENTCOM AOR to review logistics operations. This visit led them to ask whether efficiencies might be gained by combining medical warehouse distribution with nonmedical distribution in the AOR, while maintaining the quality of health care delivery. Nonmedical materiel for USCENTCOM customers is shipped primarily from the Defense Distribution Depot Kuwait, Southwest Asia (DDKS), from continental U.S. (CONUS) distribution centers—most often Defense Distribution Depot Susquehanna, PA (DDSP)—

---

[1] The studies are the "1953 Munitions Board Study of the Medical Supply System," the "1955 Hoover Commission Report," the "1965 Department of the Army Board of Inquiry on the Army Logistics System (Brown Board)," the "1965–1969 Logistics Review—U.S. Army Vietnam at the direction of LTG Mildren, Deputy Commanding General, U.S. Army Vietnam," the "1973 Bureau of Medicine and Surgery Study and Technical Workshop on Medical and Dental Supply Support," the "1985 Comptroller of the Army Installation Study," and the "1994 Department of the Army, DCSLOG Directed Analysis by the U.S. Army Logistics Evaluation Agency on Medical Logistics Policy Proponency."

and directly from vendors for certain classes of items such as food. Managed by the Defense Logistics Agency (DLA), DDKS is a contractor-owned and -operated distribution center that stores and distributes materiel in supply Classes II (such as textiles, uniforms, tents), IIIp (packaged petroleum products), IV (barrier and construction materials), and IX (repair parts). DDKS is situated in Kuwait to the north of Camp Arifjan and near the Ali Al Salem Air Base as well as the Kuwait International Airport.

U.S. Army Medical Research and Materiel Command (USAMRMC) asked RAND Arroyo Center to develop alternative options for the distribution of medical supplies in the USCENTCOM AOR and to evaluate the likely effects on cost and performance. This study investigates whether there are distribution options that would improve performance, reduce cost while maintaining current performance, or both. Although initially the study was to focus more narrowly on the question of whether stocks from USAMMC-SWA might be consolidated with those at DDKS, based upon initial analysis and in coordination with USAMRMC, we broadened the question to take a more comprehensive look at a range of options for the USCENTCOM AOR.[2]

The Current Distribution Structures for Medical and Nonmedical Supplies

Figure 1.1 shows the locations of DDKS, USAMMCE, and USAMMC-SWA.

USAMMCE

USAMMCE is a strategic distribution platform that supports three combatant commands (COCOMs): U.S. European Command (USEUCOM), U.S. Africa Command (USAFRICOM), and USCENTCOM. USAMMCE stocks approximately 13,000 unique items, or "lines," of materiel for distribution across the three AORs. In addition to providing acquisition, warehouse storage, and distribution service to provide supplies to more than 1,386 joint service customers, USAMMCE provides clinical engineering support, optical fabrication service, assembly of medical sets and kits, disassembly and reconstitution services (MESKOS), and training of customers as well as Army medical logisticians through workshops and predeployment exercises.[3] USAMMCE plays a

---

[2] In the past, questions have been raised regarding whether medical logistics should be a separate function (see Appendix A).

[3] Acquisition capabilities include processing more than 600 new item requests (NIRs) per month and direct coordination with hundreds of commercial medical vendors in multiple countries to acquire items whose life cycles may be short because of rapid changes in technology or turnover in deployed clinicians who have a preference for a specific item; workshops and training on information technology and use as well as cold chain, controlled item, and potency and dated (P&D) packing and storage protocols.

Figure 1.1
Locations of USAMMCE, USAMMC-SWA, and DDKS

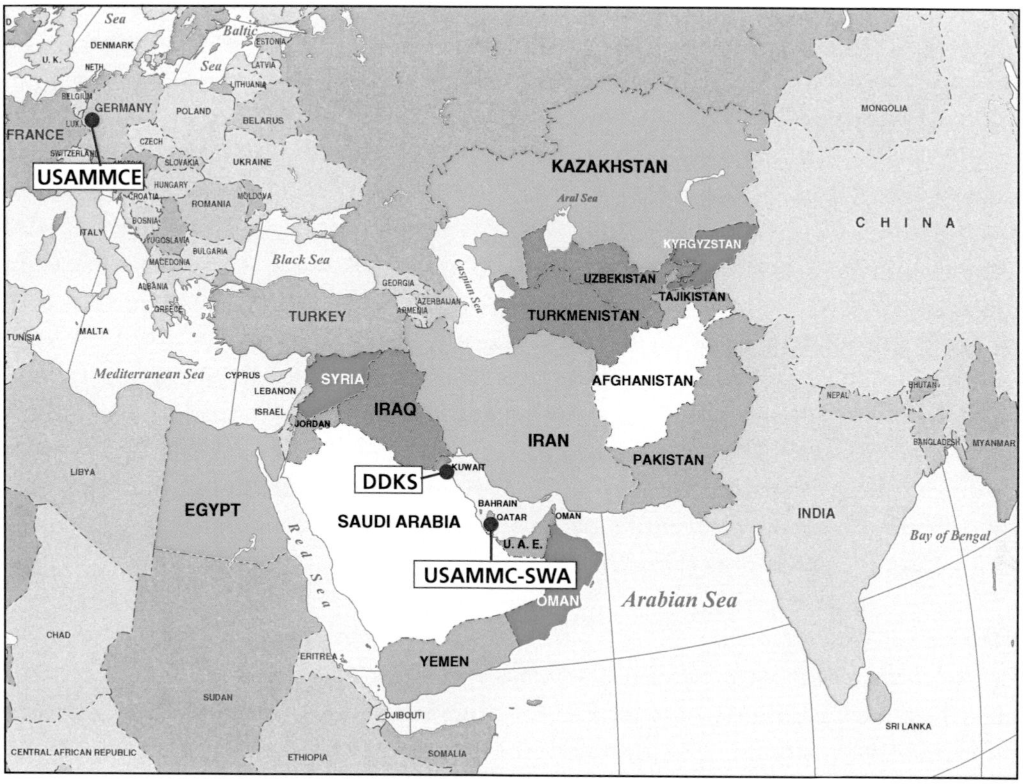

RAND *MG929-1.1*

critical role in the support of the USCENTCOM AOR in that it not only replenishes USAMCCE-SWA, but also directly fills requests for other items.

USAMMC-SWA

Prior to Operation Iraqi Freedom (OIF), what is now USAMMC-SWA was a storage location for medical unit sets in Army Prepositioned Stocks (APS). With the onset of Operation Enduring Freedom (OEF) in October 2001, Army Central Command was directed to serve as Single Integrated Medical Logistics Manager (SIMLM) with the mission to "oversee medical supplies, equipment, optical fabrication, medical gases, medical equipment maintenance and repair, and blood management efforts among all services in the theater" (Brew, 2003a) as called for by the USCENTCOM operations plan. ARCENT established a small medical logistics operation at Camp Snoopy in Qatar, where proximity to both Doha International Airport and Al Udeid Air Base provided access to strategic and intratheater air channels. In August 2002, the Army Office of the Surgeon General (OTSG), in coordination with USCENTCOM, proposed to the Army G-4 that the medical APS facility at Camp As Sayliyah, Qatar, be modified to serve as a "warm-base" distribution facility for new APS sustainment stocks. Improvements included the addition of environmental control, a vault, refrigerated and hazardous material storage, a medical maintenance workshop, office space, and high-density shelving. The task of organizing the new facility was given to the 6th Medical Logistics Management Center (6th MLMC).[4]

In the buildup to OIF, USAMMC-SWA was formally established by the 3rd Medical Command (MEDCOM) as a provisional organization at Camp As Sayliyah and subsumed the operation at Camp Snoopy in February 2003 to provide forward warehousing distribution for medical logistics and combat service support (CSS) in support of OEF and Joint Task Force Horn of Africa (Galuszka, 2006).[5] Concurrently, the Army transferred its APS medical sustainment inventory to the DLA Defense Working Capital Fund to enable reimbursable sales to all service components, and USAMMCE was designated the acquisition authority and prime vendor interface for stocks coming from CONUS to replenish USAMMC-SWA. USAMMCE maintains the direct linkage to national-level sources of supply and provides local purchase support for requirements that cannot be met through prime vendor or other Defense Supply Center Philadelphia (DSCP) acquisition programs.

DDKS

For nonmedical supplies, DLA has a forward distribution depot (FDD) in Kuwait called Defense Distribution Depot Kuwait, Southwest Asia. The DDKS facility is a contractor-owned, contractor-operated depot that stores and distributes supply Classes II (such as textiles, uniforms, tents), IIIp (packaged petroleum oil and lubrication products), IV (barrier and construction materials), and IX (repair parts). DDKS is situated in Kuwait to the north of Camp Arifjan and near the Ali Al Salem Air Base as well as the Kuwait International Airport.

In general, the purpose of a forward or regional distribution depot such as DDKS is to increase response speed, lower total distribution cost, or both. The distribution times from DDKS to Iraq and Afghanistan are similar to air shipment times from CONUS. Rather than providing a response time advantage, DDKS lowers the cost of distribution by storing large or heavy or high-volume items that are also relatively inexpensive; these items can be replenished by low-cost sealift for a small investment in additional inventory. This distribution structure provides the same response time from CONUS as compared to direct airlift, but at a fraction of the cost.

---

[4] The command and control (C2) of USAMMC-SWA is provided by the 6th MLMC, and the military management and labor is provided by deployed medical logistics units (Haddad, 2008) as well as contractor support.

[5] Though USAMMC-SWA was established as a provisional organization, there was no measure or "trigger" created by which leadership could objectively determine when it should be decommissioned.

Since DDKS became active, the Army has sought out opportunities to reduce costs while improving the performance of nonmedical stockage. Initially, USCENT-COM nonmedical theater-level sustainment stocks were stored in Army general support (GS) supply support activities (SSAs), which were stood up in early 2003. However, the Army subsequently phased out inventory in the Class II, IIIp, IV, and IX common GS SSAs.[6] In 2007, shipments from the GS SSA and DDKS were combined to improve distribution performance and transportation efficiency through larger pallets built more quickly. In 2008, process and information system changes were made so that DDKS could take over the mission of receiving and processing serviceable returns for increased warehousing and distribution efficiency.

The efficiency improvements at DDKS led naturally to the question that is the focus of this study, i.e., whether further efficiencies could be gained by also consolidating distribution of non-Class VIII and Class VIII medical supplies.

Study Methodology and Evaluation of Options

There are three steps in the research methodology used in this report. For each option, we first compared how it would affect distribution performance. Then, if the distribution performance was found to be somewhat similar or better, we analyzed how the option would affect cost. Finally, if distribution performance and cost were found favorable for the option, we examined how it would affect other capabilities.

Overview and Options Considered

The primary question to be answered in this study is whether there is a less costly medical distribution structure to support USCENTCOM that would maintain the quality of health care delivery. Five options were considered, as shown in Table 1.1. The initial options—identified from questions posed by the Joint Staff J4 and service 4s outlined in a July 2008 trip report to the USCENTCOM AOR—were to consider (1)

Table 1.1
Medical Distribution Options Considered

Option	Performance	Cost	Other
Status quo			
Prime vendor from CONUS through DDSP CCP			
Stock medical supplies at DDKS			
Consolidate at USAMMCE (no USAMMC-SWA)			
Replicate USAMMCE stocks at USAMMC-SWA			

---

[6]   A GS SSA remained in place to accept and process serviceable returns (Peltz et al., 2008).

supporting the USCENTCOM AOR with medical materiel shipped via prime vendor support from CONUS through DDSP, (2) stocking medical supplies at DDKS and closing the distribution center at USAMMC-SWA, and (3) the status quo structure. Two additional options were considered based upon preliminary data analysis and as a result of interviews with medical logisticians: (1) providing all direct support of medical materiel to USCENTCOM from USAMMCE; (2) increasing inventory at USAMMC-SWA (replicate most of USAMMCE stocks) so that it could provide most items directly to customers.

Decision Framework

We used a decision tree framework to evaluate each option (see Figure 1.2). The first criterion that each option had to satisfy was that of performance: does the option maintain or improve performance with regard to how long it takes to fill orders? The medical supply chain is focused on clinical outcomes, so timely response to needs is considered critical. Thus, performance for any proposed distribution option must be equal to or better than current performance. Second, is the option more or less costly than the status quo, with current performance considered acceptable? If an option meets these two criteria of performance and cost, then it is considered as a possible distribution option for medical materiel to the USCENTCOM AOR. In an effort to capture intangible effects of these options, we considered items such as specific geographic location and personnel training in the "other" column.

Due to the study's compressed timeline, the data collection, distribution analysis, cost analysis, and stakeholder interviews were performed concurrently. Relevant firsthand knowledge and context for the study were acquired over the course of 11 days as RAND Arroyo Center researchers (accompanied by the Medical Logistics and Operations Officer, HQDA G4/OTSG LNO, and the DSCP Medical Troop Support Planner) met with representatives of USAMMCE, USAMMC-SWA, DDKS, the

Figure 1.2
Decision Framework for Evaluating Distribution Options

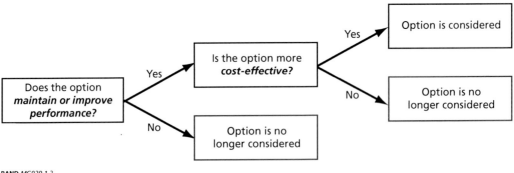

CDDOC, the 8th Expeditionary Air Mobility Squadron (EAMS) of the U.S. Air Force, the 3rd EMF of the U.S. Navy, and the Camp Arifjan Army Clinic.

Performance Measurement

To compare the performance of different options, we developed new metrics to capture, to the extent possible, the distribution flow for medical logistics. Most distribution metrics previously available to the medical community tended to focus on supply availability (e.g., fill rate), depot performance (e.g., pick/pack time), or transportation (e.g., air tender time). In this study we created a more complete picture of distribution performance in support of medical customers in the USCENTCOM AOR by integrating data from several sources. Using these data, we developed a method to measure end-to-end distribution, and in particular we sought to provide additional detail to diagnose the reasons for differences in overall distribution times.

We identified common segments of the distribution flow to compare the distribution process among options considered, as shown in Figure 1.3:

1. *Doc to receive:* The time from the initial request for materiel or the document date (*doc*) to the time that the request is received electronically at the depot (*receive*).
2. *Receive to issue:* The time that it takes for the *received* request to be prioritized and printed as a materiel release order (MRO) for the depot to *issue*.
3. *Issue to pick:* The time that it takes for the warehouse staff to identify and collect the materiel requested.
4. *Pick to triwall consolidation:* The amount of time necessary to consolidate all materiel to be packed (in a container type known as a "triwall") and shipped to a particular customer(s) or location(s).
5. *Triwall closed to arrival at destination:* The transportation time from the depot to the customer. For USAMMC-SWA shipments this final segment is divided into two:
   a. Triwall closed to Al Udeid arrival (USAMMC-SWA only)
   b. Aerial port of embarkation (APOE) arrival to aerial port of debarkation (APOD) arrival (USAMMC-SWA only).

To the extent possible, it is important that there be no "air gaps" between segments. The one element we could not capture is formal receipting by the customer, as the Theater Army Medical Management Information System (TAMMIS) does not create the equivalent of a Military Standard Requisitioning and Issue Procedures (MILSTRIP) Defense Reporting Activity (DRA) transaction signaling receipt.

Figure 1.3
Segments in the Distribution Flow of Medical Materiel

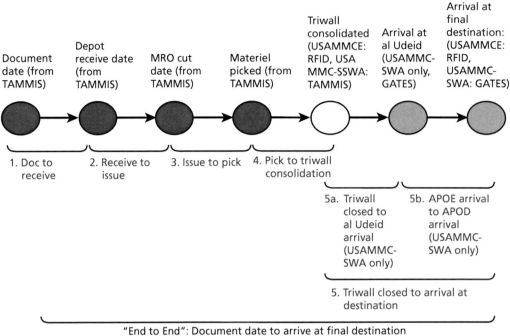

A number of data sources were used. For supply availability, depot fill rate, and depot processing (including consolidation time), we used TAMMIS system data.[7] To capture movement from depots to the customers' destinations, we used a combination of Global Air Transportation Execution System (GATES) data maintained by Air Mobility Command (AMC) and radio frequency identification data (RFID) maintained by Program Executive Office–Enterprise Information Systems (PEO-EIS). By combining these data sets at the transportation control number (TCN), we were able to track customer requisitions from their creation date back to request receipt at USAMMC-SWA or USAMMCE, determine if the requisition was backordered, then track it through materiel release order, picking, hold for consolidation, movement to the aerial port of embarkation, and transportation to the customer's aerial port of debarkation or actual physical location.

Measurements for USAMMCE and USAMMC-SWA differ somewhat based on process and transportation mode differences and on different data problems. Measuring the transportation segment for USAMMCE shipments depends largely on RFID data, as virtually all shipments go via commercial tender and so are not included in

---

7   See Chapter 6 of U.S. Army Field Manual (FM) 8-10-15 for a full description of TAMMIS.

the GATES database; nor were tender reports available to us at the TCN level.[8] In FY2008, almost all USAMMC-SWA shipments went via AMC-managed airlift and were captured in the GATES database. While we tried to supplement GATES data with RFID data for these shipments, we found relatively few cases of USAMMC-SWA shipments where we could track the RFID record from the depot to the port and beyond. This meant that for USAMMC-SWA cases we could not see the end of the consolidation process (captured at USAMMCE with the RFID tag burn date); nor could we see arrival at the customer's actual location, as our last recorded step in the process was at the customer's aerial port of debarkation.

In FY2009, USAMMC-SWA shifted a sizeable amount of its airlift to the Class VIII tender and therefore outside AMC channels. As we still had the RFID data deficits, this meant we were no longer able to follow the movement of materiel at the document level from the depot to the customer locations with the same level of detail. For FY2009 analysis, then, we had to split measurement into materiel processing at the depot and movement by the tender.

The data sources used to capture each segment of the distribution process are listed in Appendix C.

Cost Comparisons

To compare costs of transferring USAMMC-SWA operations to DDKS, we estimated how much it would cost to conduct the USAMMC-SWA distribution center mission at DDKS with the actual costs for performing this mission at USAMMC-SWA, and determined how transportation costs would change based upon the actual airlift rates from the two locations.

---

[8] RFID data have notable limitations, especially where, as in this case, we are looking for arrival at the customer's location. Many tagged shipments never show a "ping" (data record) at customer locations, whether because the tags are shielded from the interrogator, the tag battery is too weak to broadcast, or the battery is not present, among other reasons. Often a tag announces its presence for the first time long after a shipment has arrived at a destination. These inaccuracies in RFID data limit the population we can measure and force the use of decision rules about when to include and exclude data.

CHAPTER TWO

Evaluation of the DDKS and CONUS Options

In this chapter we discuss the results of our analyses of the first two options: (1) moving medical supplies to DDKS for onward distribution and (2) distributing medical materiel to the USCENTCOM AOR from CONUS through a containerization and consolidation point (CCP) at a defense depot, such as the one located in Susquehanna, Pennsylvania (DDSP). We compare performance for each option vis-à-vis the status quo. Then, if relevant (i.e., if there is a performance advantage associated with changing the status quo), we compare costs for the two options.

Background: Medical and Nonmedical Supply Chains

Before evaluating these options, it is important first to say something about the differences between the medical and nonmedical supply chains within the Army.

The Army's nonmedical supply chain is built upon major distribution centers (DCs) or strategic distribution platforms (SDPs) in CONUS, which provide direct support to OCONUS customers and are used to replenish overseas forward distribution depots. For the USCENTCOM AOR, most materiel is shipped to customers either from the SDP in Susquehanna (DDSP) or from the DLA FDD in Kuwait (DDKS). Other materiel is shipped from other DLA CONUS DCs, primarily with transshipment through the CCP at DDSP. Additionally, the small percentage of direct vendor delivery materiel (DVD) is primarily sent to the CCP for consolidation with other shipments for overseas delivery. Strategic airlift delivers materiel from CONUS directly to airfields throughout the USCENTCOM AOR, typically through commercial charters transloaded to C-17s in Incirlik, Turkey. Commercial air services and ground convoys deliver materiel from Kuwait. Additionally, some supplies are trucked or flown from the FDD in Germersheim, Germany.

In contrast to the Army's nonmedical supply chain, USAMMCE and USAMMC-SWA operate within the commercially based acquisition framework established by DLA in coordination with the Military Health System.

Currently, at the initial stage of the distribution chain, medical materiel supporting the USCENTCOM AOR is provided by medical prime vendors (PVs) and other commercial suppliers by standard electronic data interchange (EDI) transactions that ship goods directly to USAMMCE, typically to replenish inventory. The PVs present materiel for shipment to commercial carriers within 24 hours for direct delivery to USAMMCE.[1] The medical surgical prime vendor (Owens & Minor) also serves as a consolidation point for DVD orders from other suppliers, combining those into PV shipments through a program known as the medical air bridge. USAMMCE then replenishes USAMMC-SWA. Customer orders are then filled from either USAMMC-SWA, if the item is stocked there, or from USAMMCE.

CONUS

We now consider the option of distributing medical materiel to the USCENTCOM AOR from CONUS through a CCP at a defense depot such as DDSP.

The direct vendor delivery model is the closest analogue to the current medical model that relies upon prime vendor support. Beginning at some vendor location, Class IX materiel is shipped to the CCP and then forwarded on to the final APOD. Figure 2.1 compares performance for average end-to-end distribution time for shipments of Class VIII materiel using the status quo method (left) and DVD shipments through the DDSP (Class IX materiel) CCP (right).

As is evident from the figure, supporting CENTCOM from CONUS through standard channels would lead to worse support. The total time from the moment an MRO goes to the vendor for a DVD item until the materiel arrives at the customer's APOD when originating from CONUS is an average of 28 days, whereas medical end-to-end performance averages 10 days total time for Class VIII (center column). If we take out the vendor processing time for DVDs, in case medical prime vendors have better processes than the non-Class VIII DVD suppliers, the elapsed time from shipment to arrival for DVD shipments (right column) is still longer than the total Class VIII average distribution time.

Because the performance associated with the CONUS option is significantly poorer than that associated with the current medical materiel distribution structure, we did not consider this option further.

DDKS Option

We now consider the next option, that of moving medical supplies to DDKS for onward distribution. To understand this option, we need first to consider the transportation network used for medical and nonmedical supplies, and the degree to which military

---

[1]   USAMMCE serves as an OCONUS SDP.

Figure 2.1
MRO to Destination Time for USAMMC-SWA and Class IX DVDs from CONUS to CENTCOM Customers

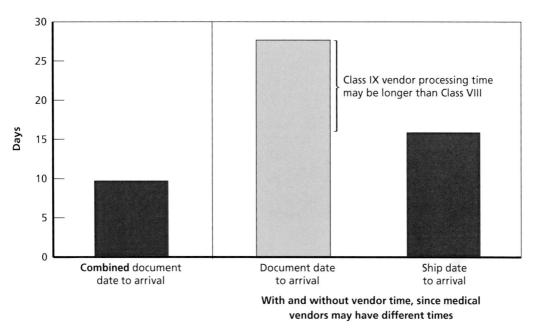

RAND *MG929-2.1*

airlift and commercial transportation are used at DDKS and USAMMC-SWA. We will then compare performance and costs for the two modes.

Background: Air Transportation for Medical and Non-Medical Materiel

As shown in Table 2.1, there are four possible types of air transportation into and within the USCENTCOM AOR. The associated cost, transportation time, and responsiveness (contingency availability) vary among them.

Table 2.1
Types of Air Transportation Within USCENTCOM AOR

Type	Use
MilAir	• C-17s and C-130s used for intra-theater airlift
Theater Express tender	• Spot bids from nine vendors serving 72 city pairs • Used by Air Mobility Division when MilAir is not available
WorldWide Express (WWX)	• Spot bids for airlift from multiple vendors
Class VIII commercial tender	• Fixed, negotiated rates from four vendors for inter- and intra-theater

MilAir. Intra-theater military airlift (MilAir) is made up primarily of C-17 and C-130 aircraft that follow regular, planned routes (channels). There is a limited supply of these "gray-tailed" military aircraft as well as the operators and maintainers necessary to keep these aircraft mission capable.[2]

Theater Express and World Wide Express. When military airlift is not available to transport materiel (of any class), there are two main commercial tender alternatives: use of the World Wide Express contract and the USCENTCOM Theater Express. Both tender agreements are negotiated and let by USTRANSCOM contracting officers and are executed through the use of a spot-bid process where the shipment details (where to, by when, how much weight, what type of material, etc.) are released to the contracts' vendors, who in turn reply with a bid to move that particular shipment. The World Wide Express tender (for which four vendors are currently active bidders with contracts) is targeted toward the worldwide transportation of subpallet shipments (e.g., smaller cubed volume, lower-weight items).[3] In contrast, transportation via the Theater Express tender covers all shapes and sizes of shipments but is strictly limited to 72 city-pairs (must be able to move from one city in the list to any other) in the USCENT-COM AOR.[4] Any materiel arriving at Al Udeid Air Base in Qatar with a movement requirement of less than 72 hours, and without an immediately available organic channel flight, is put out for spot bid under Theater Express.

Although there are nine primary vendors for the Theater Express tender, it is not guaranteed that each shipment offered for movement under this tender will be bid upon by all nine or even any right away. In practice, at the 8th EAMS at Al Udeid, there are occasions where pallets will sit for long periods of time waiting for additional shipments to a particular location prior to any bids being offered by the commercial vendors.

Class VIII Commercial Tender. Due to the time-sensitivity of many Class VIII items (because of both the perishable nature of some items and the immediate health needs at the end-user site), USTRANSCOM negotiates and manages a separate Class VIII tender with four commercial vendors. This tender has four distinct rate categories for standard Class VIII shipments (medical commodity items), cold chain items,[5] dangerous goods (hazardous materials), and life-and-death shipments (which reduces the 96-hour delivery requirement shared by the previous three down to 48 hours or less).

---

[2] During our research, the 8th EAMS at Al Udeid Air Base and the Tanker Airlift Control Center at Scott Air Force Base, Illinois, stated that C-17s are loaned to the AOR and can be pulled at any time; while C-130s are dedicated to the theater commander but are used for personnel transport as well as transport of supply. Therefore, the supply of organic military aircraft is considered limited.

[3] WWX is available for use by medical logisticians, but it is not typically used due to the existence of the specialized Class VIII tender.

[4] See Appendix I for a listing of Theater Express cities included in the 72 city-pairs.

[5] These are medical items that must be kept cold throughout the shipping and handling process.

The option of fulfilling medical materiel demands from DDKS requires some discussion of the degree to which MilAir and commercial transportation are used at DDKS and USAMMC-SWA, as shown in Figure 2.2. The airlift mix used from both USAMMC-SWA and DDKS in FY2009 is different from how each of these organizations operated in FY2008. Given our need to project forward, for our analysis we focus on FY2009 processes, data, and performance.

As shown in the top half of the figure, approximately two-thirds of DDKS shipments by weight are being sent via commercial tender pallet using the USCENTCOM Theater Express tender, while one-third is being shipped via organic military air assets. In comparison, as shown in the bottom half of the figure, USAMMC-SWA utilizes the Class VIII tender for roughly 40 percent of shipments, by weight, and uses military-managed airlift (66 percent Theater Express and 33 percent organic) for the remaining 60 percent.[6]

Figure 2.2
Relative Use of Air Transportation Modes for DDKS and USAMMC-SWA

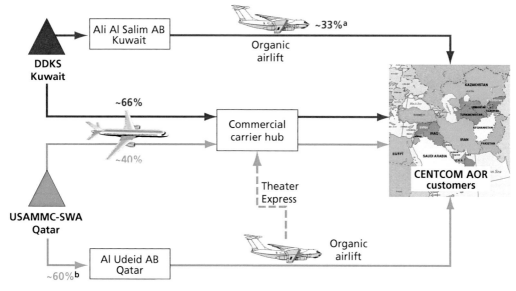

[a] Based upon TACC tracked AOR actuals for Theater Express.
[b] USAMMC-SWA ratio of tender to total rose in January–February 2009 to ~55%.
RAND MG929-2.2

---

[6] In FY2009, the Chief of Support Operations at USAMMC-SWA began utilizing the same Class VIII commercial tender contracts that were originally written for USAMMCE. This was made possible through a change in the contract in August 2008. The Class VIII tender agreement was written to allow materiel to originate out of Germany and Qatar, and it was updated in February 2009 to allow materiel to originate out of Iraq and Afghanistan. The tender is currently under revision to include shipments to Yemen, Pakistan, and all of USAFRICOM; these modifications will be included when the tender is renewed in August 2009.

In FY2009, USTRANSCOM made a major push to use the Theater Express commercial tender for airlift out of Kuwait (to include DDKS), due to restrictions by the Kuwaiti government on the number of military airlift that can be on the ground along with available capacity at Ali Al Salem Air Base and Kuwait International Airport. Significant delays existed in the past due to the high volume moving through Kuwait and the limited capacity due to the military airlift restrictions. The increase in the use of Theater Express acts in a sense to lessen these aerial port constraints.

USAMMC-SWA has also increased the use of Class VIII commercial tenders, although this increase has been more of a fine-tuning action since there are no constraints on the use of military aircraft at Al Udeid Air Base. Some outlying locations in Iraq and Afghanistan can be reached more quickly via tender compared to military air. Other places that are central hubs or high-volume locations, such as Bagram, Baghdad, and Balad, are very straightforward to get to via organic military airlift, since there are regularly scheduled channel flights to these locations.

Performance

Figure 2.3 compares transportation times for military airlift in FY2008 (dark blue bars) and Class VIII commercial tender in FY2009 (light blue bars). The MilAir transportation times are similar to those of the Class VIII commercial tender for destinations that receive a lot of air cargo traffic such as Al Asad and Bagram. However, for all other destinations, the Class VIII tender has a clear transportation performance advantage.

The reason for the dramatic difference in times rests in the difference of the operating policies of the two organizations providing air transportation service. The MilAir routes are managed to maintain high utilization of aircraft on each route; if volumes drop on a particular route, then service frequency may also be dropped. However, the Class VIII commercial tender aircraft originate travel every day from both Frankfurt and Doha en route to Sharjah, United Arab Emirates, where cargo (from both USAMMCE and USAMMC-SWA) is consolidated for transfer to aircraft serving specific locations within the USCENTCOM AOR. The Class VIII tenders are not managed to maximize aircraft utilization but rather to hit service targets, and the cost of maintaining the service level is passed to the customer in the price of the tender.

We note that although both DDKS and USAMMC-SWA have increased the relative use of tender, the extent of use of the Class VIII tender by USAMMC-SWA is directly managed and controlled by the USAMMC-SWA staff and leadership, whereas the increase in Theater Express usage by DDKS is dictated by the Air Mobility Division (AMD) of the Combined Air Operations Center (CAOC) at Al Udeid Air Base, Qatar.[7]

---

[7] DDKS management does not determine the airlift mix for items distributed out of the depot.

Figure 2.3
Comparison of Transportation Times for Military Aircraft and Class VIII Commercial Tender

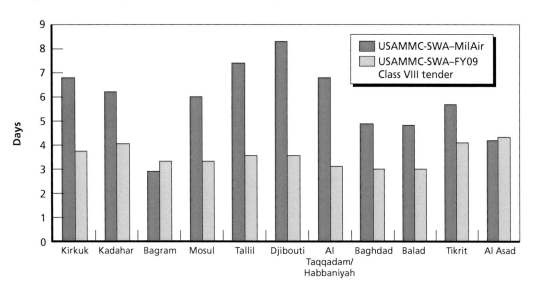

NOTES: Class VIII tender data: October–December 2008, USTRANSCOM Class VIII tender payment and performance tracking. MilAir data source: June–September 2008, USAMCC-SWA production reports.
RAND *MG929-2.3*

Figure 2.4 shows the change in overall times between FY2008 and FY2009 for the two sources, along with average times for each major segment of the overall process. In FY2009, both DDKS and USAMMC-SWA changed their mix of transportation usage, resulting in better end-to-end performance for both locations. As discussed earlier, the AMD has made more substantial use of the Theater Express tender service to distribute items out of DDKS, especially in support of Afghanistan customers, and USAMMC-SWA has increased its use of the Class VIII tender, especially to reach customers not served as well by organic military air. The end result is that USAMMC-SWA has retained its performance advantage over DDKS, though the gap has narrowed slightly. Times from MRO to arrival at the APOD from DDKS averaged about 6.5 days in FY2009, with the average at about 4 days for shipments from USAMMC-SWA. The advantage for USAMMC-SWA lies primarily in the MRO to pick segment, with some advantage also in the transportation segment.

For USAMMC-SWA, we see a change in transportation time, which on average declined from almost 3 days to just over 2 days between FY2008 and FY2009. The change resulted primarily from reducing the outlier times USAMMC-SWA had experienced in reaching smaller, out-of-the-way customers in locations like Tallil and Al Asad, while it continues to use the robust organic military air network to reach Balad, Baghdad, and the like directly from Al Udeid Air Base.

Figure 2.4
Average Segment Times: DDKS and USAMMC-SWA

- In FY09, both DDKS and USAMMC-SWA changed transportation practices to include use of Theater Express commercial tenders (DDKS) and Class VIII commercial tenders (USAMMC-SWA)

- Analysis of FY09 data show that USAMMC-SWA still has an advantage over DDKS

  — USAMMC-SWA transportation performance remains faster for delivery to Afghanistan and outlying regions in Iraq (e.g., Kirkuk, Mosul, Tallil)

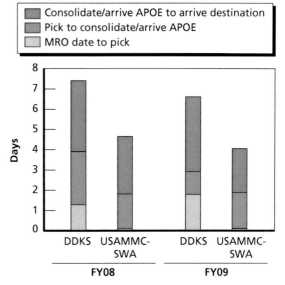

SOURCES: DDKS: Strategic distribution database; USAMMC-SWA: TAMMIS and GATES.
RAND MG929-2.4

In comparison, the use of the Theater Express tender to distribute materiel from DDKS has not resulted in faster transportation times, as seen above, but has led to a significant reduction in the amount of time cargo is held before it is delivered to the airlifter. At DDKS, pallets may be held at the depot until called forward to the APOEs serving military flights, either at the Ali Al Salem Air Base or the part of Kuwait International Airport used for military flights (including commercial charters managed by the AMD). Both locations are constrained with respect to the number of airframes they can accommodate and the number of pallets they can store awaiting lift. Theater Express flights originate primarily from the commercial section of Kuwait International Airport, which is less constrained. Because DDKS is now less obliged to hold pallets awaiting airlift, this part of the overall process saw a drop from about 2.5 days to just over 1 day between FY2008 and FY2009.

Figure 2.5 shows FY2009 combined consolidation and transportation times (equivalent to the last two time segments or pick to arrive at destination in Figure 2.4) from DDKS and USAMMC-SWA to specific locations in the USCENTCOM AOR of specific interest to USAMRMC.[8] We show the combined segment times, since the use of Theater Express for DDKS shipments actually improved the pick to arrive at the

---

[8] October through December 2008 for origin to APOD pairs where there were a minimum of ten observations. See Appendix G for a list of all origin-destination pairs used in the analysis.

Figure 2.5
Pick to APOE Arrival Times from DDKS and USAMMC-SWA to Specific Locations in the USCENTCOM AOR

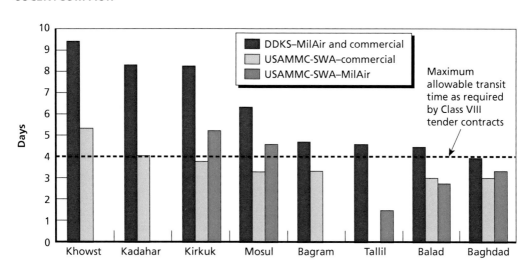

NOTES: October–December 2008 transportation times from origin to APOD where there were a minimum of 10 observations. USAMMCE and USAMMC-SWA commercial air data source: USTRANSCOM. USAMMC-SWA MilAir data source: USAMMC-SWA production reports. DDKS: Strategic Distribution database.
RAND *MG929-2.5*

APOE segment rather than the actual transportation time. The figure shows that times from USAMMC-SWA are faster than those from DDKS for all of these locations.[9] The red columns at the left of each set show the average times using a combination of MilAir and commercial tender from DDKS. The dark blue columns at the right of each set show the average times for the combination of MilAir and Class VIII commercial tender from USAMMC-SWA. The light blue columns in the middle of each set show the times for only the Class VIII commercial tenders originating from USAMMC-SWA. For reference, the horizontal dotted line shows the maximum allowable transit time required by the Class VIII tender contracts.[10]

Costs

Now that we have seen the relative performance of USAMMC-SWA and DDKS, we compare the cost of the USAMMC-SWA organization performing its distribution

---

[9] USAMMCE and USAMMC-SWA commercial air data source: USTRANSCOM; USAMMC-SWA MilAir data source: USAMMC-SWA Production Reports; DDKS transportation data source: Strategic Distribution Database.

[10] The order of bars in the chart does not represent the relative importance of one destination versus another nor the aggregate weight delivered to any one location. In fact, since so much weight is shipped to Balad and Baghdad, the average performance charted in Figure 2.4 reflects those two cities' columns in Figure 2.5 more than any others.

Table 2.2
Data Used for Comparison of USAMMC-SWA and DDKS Costs

USAMMC-SWA	DDKS (Agility PWC)
• Military manning and contractor personnel details and costs* • Equipment and facilities costs • MilAir and Theater Express tender usage from GATES, SWA Production Reports, and billed cost actuals/tariffs from USTRANSCOM • Billed cost actuals and performance data for the Class VIII tender contract by USAMMC-SWA along with the vendor contracts and rate tables • Facility schematics and utilization rates • Requirements for handling, and capacity for specialty medical materials	• Contract costs contained in the DDKS contract (SP3100-05-C-0020) with Agility PWC across first four contract years • Government personnel costs • MilAir and Theater Express Tender usage from GATES and billed cost actuals/tariffs from USTRANSCOM • Facility schematics and utilization rates • Analysis of available specialty handling capability for handling cold chin, hazardous materials, controlled narcotics, etc.

*SWA annual total warehousing personnel costs do not include personnel in special functions who would have to transfer to DDKS.

center mission to the estimated cost for DDKS performing the USAMMC-SWA mission. This comparison required various sources of data, as shown in Table 2.2.

We utilized the following information to detail the cost of USAMMC-SWA operations:

- Composite Standard Pay and Reimbursement Rates for military personnel.[11]
- EAGLE Contract W52P1J-08-C-0016 detailing contractor support to operations at USAMMC-SWA.
- Cost of maintenance and operation of equipment and facilities at USAMMC-SWA.[12]
- Organic military airlift usage from the GATES system.
- TWCF rates for organic military airlift from Qatar as reported by USTRANSCOM on a fiscal year basis.
- Theater Express utilization and rates as tracked by USTRANSCOM/J8.
- Billed costs for Class VIII tender as reported to USTRANSCOM/J8 by DFAS.[13]
- Facility schematics and utilization rates, to include the requirements and capacity for the handling and care of specialty medical materiel.[14]

---

[11] The composite rates are issued by the Office of the Under Secretary of Defense (Comptroller) for use in studies and analysis.

[12] Information was collected during the site visit from military leadership at USAMMC-SWA.

[13] See Appendix B for a detail of how Class VIII invoices are processed.

[14] This information was collected regarding USAMMC-SWA so as to accurately determine costs associated with the replication of equal capability at another location. Specialty medical materiel includes cold chain items, narcotics, pharmaceuticals, and hazardous materials.

Information required from DDKS included the following:

- Composite Standard Pay and Reimbursement Rates for military personnel and U.S. government civilians.
- Agility PWC Contract SP3100-05-C-0020 that contains the costs for operation of DDKS.
- Facility schematics and utilization rates, to include the capability and capacity for the handling and care of specialty materiel.
- Organic military airlift usage from the GATES system.
- Transportation Working Capital Fund (TWCF) rates for organic military airlift from Kuwait as reported by USTRANSCOM on a fiscal year basis.
- Theater Express utilization and rates as tracked by USTRANSCOM/J8.

All cost data, where possible, were taken from actual billed costs. For the active-duty military members or the government civilians, we used the FY2008 Department of Defense (DoD) Military Personnel Composite Standard Pay and Reimbursement Rates tables to assign the annual cost for each member as has been directed by the DoD for the purpose of studies and analysis. We extracted the costs of contracted personnel and local nationals directly from the contracts and any applicable contract modifications on record.

Although we collected the costs of the entire USAMMC-SWA operation (to include medical maintenance, optical fabrication, etc.), the cost analysis is limited to the portion of the operation that, in this option, would be relocated to DDKS and executed by the same type of workforce already there: the stocking, picking, and distribution activities. The FY2008 cost of these USAMMC-SWA warehousing functions plus the specialty medical personnel was roughly $5.23 million. In moving the USAMMC-SWA mission to DDKS, the upper bound of the distribution center operational efficiency that could possibly be gained is equal to the amount of the workload that could be absorbed or performed more efficiently at DDKS. Since $1.41 million in annual USAMMC-SWA costs would have to transfer to DDKS to perform specialty medical functions for which DDKS does not have current capability, the operational efficiency upper bound is equal to $3.82 million for FY2008 (or the $5.23 million to operate USAMMC-SWA less the cost of specialty medical personnel).

Table 2.3 lists the FY2008 billed costs for USAMMC-SWA and DDKS, as extracted from the sources mentioned earlier. The periods of performance for the EAGLE Contract at USAMMC-SWA (annual) and the Agility PWC Contract at DDKS (annual) drove the decision to use one year as the time basis for this comparison. We used data from FY2008.

Table 2.3
Cost Actuals for Personnel, Facilities, and Equipment Used as Comparison Baseline

	USAMMC-SWA	DDKS[f]
Class of materiel	VIII	II, IIIp, IV, IX
Cost of personnel (FY2008)	$5.23 million	$12.14 million
Cost of equipment (FY2008)	$0.23 million	$7.66 million
Additional personnel necessary at DDKS[a]		$1.41 million
Additional contract costs (FY2008)[b]		$28.59 million
New facilities cost[c]		$6.53 million
Number of pounds moved (FY2008)[d]	4,981,018	215,799,932
Number of issues (FY2008)[e]	180,451	1,254,741

[a] Additional personnel billets that must be filled at DDKS in order to successfully and efficiently order, manage, handle, treat, and ship Class VIII specialty items, such as pharmaceuticals (represents an upper bound for the inclusion of additional personnel).

[b] Additional contract costs incurred include cost per issue over 1 million issues, covered square footage over 1 million square feet, uncovered square footage over 1 million square feet, employee overtime, etc.

[c] Cost of construction necessary at DDKS to absorb the SWA mission of 985 sq. ft. of refrigerated storage (at an estimated cost of $125/sq. ft.) and 54,105 square feet of covered warehouse storage (at an estimated cost of $100/ square foot) assuming DDKS is operating at 100 percent utilization (represents an upper bound for the facilities cost).

[d] Total weight (pounds) moved indeterminate of delivery shipping method (ground, air, or sea).

[e] Total number of issues completed at a depot indeterminate of delivery shipping method (ground, air, or sea).

[f] Assumes that the inclusion of Class VIII materiel at DDKS does not fundamentally alter the cost structure of the current DDKS contract with Agility PWC.

For USAMMC-SWA, the cost of personnel represents the cost of the active-duty military ($2.57 million) and contractor personnel (EAGLE Contract at $2.66 million) responsible for warehousing and distribution. USAMMC-SWA personnel stated a value of $0.23 million for the maintenance and operating costs of equipment at the depot. There were no additional contract costs, nor additional facilities costs at USAMMC-SWA to include in this analysis. While there are plans to relocate USAMMC-SWA from Camp As Sayliyah to the adjacent Al Udeid Air Base, this analysis assumes that the move would not affect the costs of the operation.

To determine what the distribution center costs for handling additional volume at DDKS would be, we needed the warehouse operating costs and volume at DDKS. For DDKS, the cost of personnel is the sum of the U.S. government individuals ($2.59 million) and the Agility PWC contract personnel ($9.56 million). The contract held by Agility PWC to operate DDKS was originally written as a one-year contract with four option years. FY2008 represented the third contract year (second exercised option). Due to an increase in covered as well as uncovered square footage, additional contract costs were added to the contract for FY2008. The cost of the additional square footage along with an assessed cost per issue (for each issue in excess of one million per year) and employee overtime are incorporated in the additional contract costs in FY2008 of $28.59 million.

The FY2008 cost of equipment and facilities was $7.66 million, according to the DDKS contract. An overarching assumption for all aspects of analysis concerning the Agility PWC contract for operating DDKS is that the inclusion of Class VIII materiel at DDKS does not fundamentally alter the cost structure of the current contract. For example, the cost of picking and packing Class VIII materiel is assumed to be on par with the picking and packing for the classes of materiel currently managed at DDKS: II, IIIp, IV, and IX.

To handle the new mission, Class VIII specialty personnel would have to transfer from USAMMC-SWA to DDKS. These billets would have to be filled at DDKS in order to successfully and efficiently order, manage, handle, treat, and ship Class VIII specialty items, such as pharmaceuticals. The cost of the additional personnel necessary at DDKS is $1.41 million (cost in FY2008 at USAMMC-SWA).

For the purposes of adding the Class VIII mission to DDKS, we assumed two cases: one where the equivalent of the USAMMC-SWA warehouse space would have to be added and the other where this space requirement could be absorbed into the current DDKS facilities. The case in which the new facilities cost is included would apply if DDKS were operating at or near 100 percent effective utilization.[15] The associated facilities cost of adding the Class VIII mission to DDKS would likely fall in between these two bounds. Due to the terms of the Agility PWC contract, the U.S. government would likely be billed at some fee-for-service cost rather than being charged the cost of construction if construction were necessary. However, for this analysis, we assume the new facilities cost ($6.53 million) to be borne in the first year but spread across all DDKS issues, medical and nonmedical.[16] The $6.53 million is the estimated construction cost for 985 square feet of refrigerated storage and 64,015 square feet of covered, enclosed warehouse storage.[17]

The number of pounds moved and number of issues are the FY2008 activity levels. We used FY2008 data for the measurement of these factors to match the period of the cost data. In FY2008, DDKS shipped 215.80 million pounds through 1.25 million issues and USAMMC-SWA shipped 4.98 million pounds through 180,000

---

[15] DLA verified that the DDKS facility is operating at full capacity in accordance with its own policy.

[16] The cost of new facilities is included as a lump sum due to the nature of the contract with Agility PWC, in so far as that Agility PWC owns the land and facilities and does not consider the depreciation schedule for these assets when charging the U.S. government for the use of the assets. This assumption is consistent with the cost structure used for the inclusion of additional indoor and outdoor square footage that has been added to DDKS across the course of the contract.

[17] The cost estimates used for refrigerated storage and covered, enclosed warehouse storage were estimated to be $125 per square foot and $100 per square foot, respectively. These estimates were derived using an average of both military and commercial construction project costs worldwide.

issues. These values were used to normalize the DDKS operation to estimate the cost of performing the USAMMC-SWA mission under the DDKS current cost structure.

To compare costs, we estimated how much it would cost to conduct the USAMMC-SWA distribution center mission at DDKS with the actual costs for performing this mission at USAMMC-SWA, and determined how transportation costs would change based upon the actual airlift rates from the two locations. The cost difference is portrayed by the equation at the bottom of Figure 2.6.

We combined the DDKS cost and volume data to develop cost factors, or rates, to estimate the DDKS operating costs of performing the USAMMC-SWA mission using two different factors: the cost per pound at DDKS ($/pound) and the cost per issue ($/issue). We chose these two factors in order to eliminate the bias that might favor DDKS if we were to simply use weight moved (because DDKS moves much heavier items than the Class VIII supplies moved by USAMMC-SWA) or the bias that might favor USAMMC-SWA if we were simply to use the number of issues processed (because the items that USAMMC-SWA moves are typically very small). These rates

Figure 2.6
Comparable Cost Analysis Methodology: USAMMC-SWA Compared to DDKS

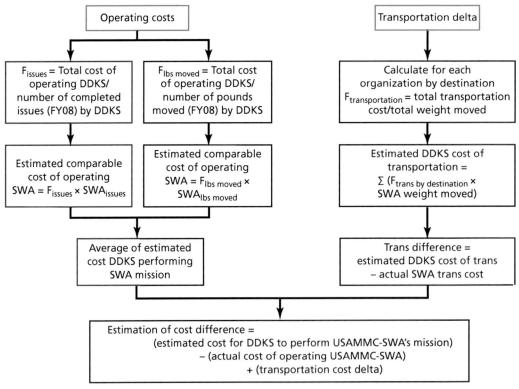

were multiplied by the FY2008 USAMMC-SWA volume to get estimated DDKS costs for performing this mission. We then took the average of the two estimated costs to estimate the cost of performing the USAMMC-SWA distribution center mission at DDKS.

As we are not able to conclusively state whether or not Agility PWC has the additional capacity at DDKS to absorb the USAMMC-SWA Class VIII mission, we performed the cost calculations for two cases: one where no new construction is required and another where new construction is necessary. The USAMMC-SWA operating cost variable in the equation is populated using the actual cost data, which is dominated by the cost of personnel.

To calculate the transportation difference (or "transportation cost delta"), the weight moved by USAMMC-SWA was decomposed by country (Iraq and Afghanistan) by month. We assumed that any materiel shipped by Class VIII tender would have to continue being shipped via this mode due to special handling requirements. Thus, we applied the USAMMC-SWA airlift mix (Class VIII tender (40 percent), organic airlift (20 percent), and Theater Express tender (40 percent)) using the DDKS rates, by destination country, with the USAMMC-SWA volume by weight so as to accurately estimate the cost of delivery if the items were to have been supplied direct from DDKS. Finally, we determined the difference between the USAMMC-SWA transportation costs versus the estimated delivery costs from DDKS. Note that for the transportation rate portion of the analysis we assumed that the replenishment cost would be equal to USAMMC-SWA and DDKS from USAMMCE via the Class VIII tender.

The equation shown at the bottom of Figure 2.6 was used to determine the relative costs of using DDKS and USAMMC-SWA for performing the USAMMC-SWA mission. Figure 2.7 shows the results of this analysis.

In the left-hand graph, the left blue column shows the absolute costs of USAMMC-SWA monthly operation, and the right column in red shows what we would project it would cost for DDKS to perform the SWA mission on a monthly basis. The dark red portion of the bar represents the aforementioned effect of additional construction, in essence establishing the upper bound of the monthly cost to perform the USAMMC-SWA mission from DDKS. During the project, DDKS government management as well as Agility PWC leadership stated that DDKS is operating at full utilization, but when solicited, neither DLA nor Agility PWC provided definitive information on whether additional construction would be needed for this mission. Through visual observation while visiting DDKS and through RAND Arroyo Center's other work involving DDKS, we know that there is some possibility for manipulations of space usage within the DDKS footprint that might yield the space necessary to absorb the USAMMC-SWA footprint. Thus, we bound the costs with and without construction. Any change to the current structure would definitely require negotiation and could possibly change the cost structure of the contract governing the facility.

Figure 2.7
Costs of Performing USAMMC-SWA Mission Through DDKS

The operating cost to perform the USAMMC-SWA mission would likely be a little more if done through DDKS; transportation costs are similar.

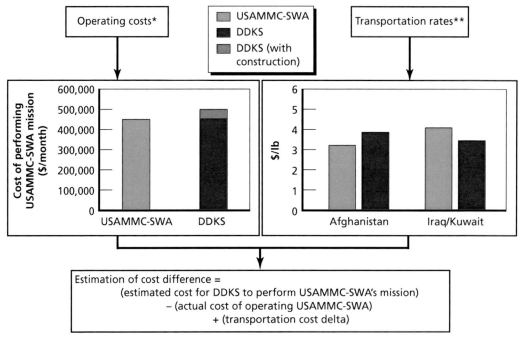

*Monthly operating costs determined using FY08 annual data , based upon transaction costs at each location.

**Transportation rates using October–December 2008 data; rates are not at steady state as USAMMC-SWA ratio of tender to total continues to increase.

RAND *MG929-2.7*

In the right-hand graph, the columns show the aggregate transportation rates from October through December 2008 to locations in Afghanistan, Iraq, and Kuwait via military airlift, military contract airlift (Theater Express), and Class VIII tender airlift. Forty percent of the airlift out of both depots is assumed to be via the Class VIII tender to match what was used by USAMMC-SWA during this same period. This is supported by our discussions with the military officers at USAMMC-SWA. Their decisions on the use of the Class VIII tender have been driven by the need for special handling and strict, quick delivery time requirements necessitated by specific types of materiel (e.g., cold chain, narcotics, etc.). The cost of using the Class VIII tender from Kuwait is assumed to be equal to the current costs out of Qatar, which seems reasonable based on our discussions with the tender managers at USTRANSCOM as well as the tender vendor representatives from National Air Cargo and United Parcel Service. The remaining 60 percent is calculated using a two-thirds, one-third mix between

Theater Express rates and organic military airlift rates, respectively. This mix was determined using actual transportation data for the Theater Express and the organic military airlift weight percentage for USCENTCOM quoted by the Tanker Airlift Control Center (TACC) within the U.S. Air Force's Air Mobility Command. These transportation rates and mixes of airlift providers directly affect the relative costs for shipments from the two depots displayed in this graph and the following charts. Again, for the transportation rate portion of the analysis we assumed that the replenishment costs would be equal whether to USAMMC-SWA or to DDKS from USAMMCE via the Class VIII tender.

We also conducted sensitivity analysis on the cost analysis to determine the effect on costs of a shift in operations toward Afghanistan. To do this analysis, we increased the number of pounds of Class VIII materiel shipped to Afghanistan and decreased the Class VIII pounds shipped to Iraq, in accordance with the FY2009 trends and planning. Both the USAMMC-SWA monthly operating cost and the estimated cost to perform the USAMMC-SWA mission on a monthly basis from DDKS are constants in the cost analysis. The "transportation cost delta" variable in the cost equation drives changes in relative costs. The difference in transportation rates, to Iraq and to Afghanistan, from each depot applied to the weight moved out of USAMMC-SWA to these locations causes the variation.

The two plots in Figure 2.8 show the effect of the transportation rates to Iraq to Afghanistan on the overall relative costs as the mix of shipments to the two locations varies. A bar displaying value to the right of the center vertical axis indicates that there is a cost advantage to continuing to supply USCENTCOM Class VIII materiel from USAMMC-SWA. Conversely, a bar pointing to the left indicates that performing the USCENTCOM Class VIII mission from DDKS would generate a cost savings over continuing the mission from USAMMC-SWA.

The baseline case ("current") represents the average monthly Class VIII weights moved in October to December 2008. We assumed a direct, linear relationship between troop levels and Class VIII materiel requirements so that a change in troop levels in a particular country would lead to an equal change in Class VIII weight required to be moved to that country. For each set of conditions, there are two bars plotted, blue and red. The blue bars correspond to the operating cost estimates that assume no construction costs are necessary at DDKS for Class VIII mission absorption. The red bars represent the cases where the cost of new construction is included in the estimate for monthly DDKS operating rates.

Comparing the top plot to the lower, it is evident that there is a greater effect from decreasing the weight shipped to Iraq versus increasing the weight shipped to Afghanistan. The top plot shows that a decrease in weight shipped to Iraq, while holding weight to Afghanistan constant, drives the current DDKS cost advantage to shift to a cost advantage favoring USAMMC-SWA. Increasing Afghanistan, while holding Iraq constant, has a less dramatic impact on the cost comparison, as is shown in the lower plot.

Figure 2.8
Transportation Costs with (A) Decrease in Shipments to Iraq, (B) Increase in Shipments to Afghanistan

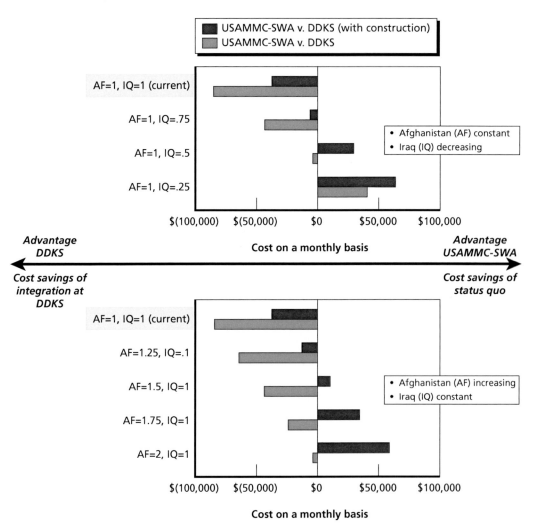

RAND *MG929-2.8*

Figure 2.9 combines the increases in Afghanistan and decreases in Iraq. For the cases in which troop levels in Afghanistan increase to 1.5 times the early FY2009 level and Iraq levels decrease in 25-percent steps from the early FY2009 level, the DDKS cost advantage shifts to a cost advantage favoring USAMMC-SWA. When the Afghanistan weight is doubled and Iraq is at one-quarter, the analysis shows a cost advantage that supports continuing operations from USAMMC-SWA.

Figure 2.9
Combined Effect on Cost of Shifting Focus to Afghanistan and Away from Iraq

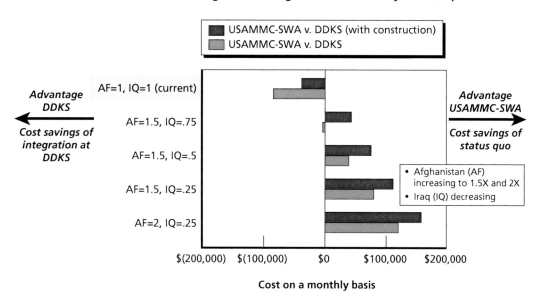

RAND *MG929-2.9*

In all of the sensitivity analysis cases, the advantage is of limited significance, as it is less than $160,000 per month. Therefore, as with performance, a clear cost rationale does not exist to support the movement of the USAMMC-SWA mission to DDKS.[18]

Conclusion

Our analysis does not make a case for moving USAMMC-SWA stocks to DDKS. Neither the FY2008 nor the FY2009 data indicate that DDKS would offer equal or better performance. The cost analysis concludes that with the drawdown of forces in Iraq and the increase in Afghanistan, a potential cost advantage for conducting medical distribution operations from DDKS disappears and with it any cost rationale for altering the current operations. There are also uncertainties due to the Agility PWC contract that we were not able to estimate. These include: What is the cost of merging the IT systems? What is the cost of treating and merging the cold chain items, and how would that affect the Agility PWC contract? Also, should any consideration be given to the fact that the Agility PWC contract expires in 18 months? These uncertainties could be further investigated, but we did not focus on them since the DDKS option does not appear to offer a clear advantage.

---

[18] We completed the cost analysis in parallel with the performance analysis due to the timing of the project and the fact that the performance difference could not be made clear until there were sufficient FY2009 data to analyze.

CHAPTER THREE

Evaluation of Options to Support the USCENTCOM AOR Class VIII Requirements from One Location

In this chapter we focus on the two remaining options: consolidation of USAMMC-SWA operations at USAMMCE, and replication of USAMMCE at USAMMC-SWA. Each of these options offers a solution where Class VIII supplies to the USCENTCOM AOR are sourced from one primary location.

Consolidation at One Location

We will then look at the improvement in distribution performance that could be attainable were the two distribution platforms consolidated at one location and follow this discussion with an analysis of transportation times to the USCENTCOM AOR from USAMMCE and USAMMC-SWA. Next, we compare the cost associated with consolidating medical materiel distribution support for the USCENTCOM AOR at USAMMCE. Finally, we discuss some of the potential impacts with regard to capabilities at USAMMC-SWA other than materiel distribution.

We will first compare distribution times when deliveries originate at USAMMCE and USAMMC-SWA. Consolidation of USAMMC-SWA operations at USAMMCE requires consideration of the performance and cost implications as well as the impact on other activities that constitute a portfolio of capabilities that are forward in Qatar.

Performance

We analyzed the distribution process performance for both USAMMCE and USAMMC-SWA based upon FY2008 TAMMIS, RFID, and GATES data. Our analysis focused on the top 20 customers across 10 destinations in USCENTCOM (see Appendix D for the list of the customers used for the analysis). In order to be included in the analysis a transaction must have had a "termination point" such that the requisition was complete and closed.

Before comparing distribution performance from USAMMCE and USAMMC-SWA, it is important to review the support structure for the theater, since this structure directly affects direct customer support from the two locations. Figure 3.1 shows the current requisition and supply flows for the theater.

USAMMC-SWA is the primary and first source of materiel for customers in the AOR, with USAMMCE playing the secondary direct support role but the primary supplier interface and theater replenishment role. If a medical organization in the USCENTCOM AOR needs an item, it enters a requisition into TAMMIS. If the item is not available locally, the requisition is electronically passed back to USAMMC-SWA. If the item is not in inventory there, the requisition is then passed back to USAMMCE via a batch process. USAMMCE then fills the requisition if the item is in inventory there, which is the case most of the time. If USAMMCE does not have the item in stock, though, then it is backordered for delivery to USAMMCE, which will then ship it to the customer. USAMMCE orders all items from suppliers for the USCENTCOM AOR, as well as for the USEUCOM and USAFRICOM AORs, and it replenishes USAMMC-SWA.

USAMMC-SWA is operated by a combination of contract and military personnel. Through this combination, it operates seven days per week. USAMMCE has stayed on a five-day per week schedule with its local civilian workforce throughout OIF and OEF.

Figure 3.1
Current Requisition and Supply Flows

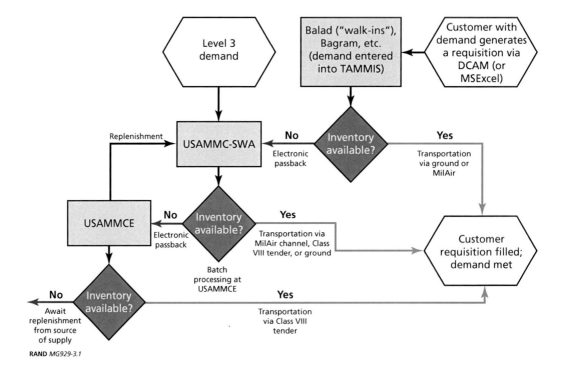

It has previously been reported that distribution times from USAMMC-SWA are faster than times from USAMMCE. This has been used to support the value and criticality of having inventory at USAMMC-SWA.[1] However, what has not previously been appreciated is that the driving factor in the difference is the structure of the system described above and not anything specific to the respective processes at the two locations. With a change in the support structure that would make USAMMCE the primary customer support option, its support would be about the same in terms of performance as that provided by USAMMC-SWA.

We are able to show this through the new combination of data sources, which enables greater measurement fidelity of the system.[2] Figure 3.2 shows the time it takes to complete the first step in the distribution process from the document creation at the customer level until arrival at the depot that fills the order. The height of the lower column segment in black shows the median or 50th percentile time, the top of the

Figure 3.2
Comparison of Times from Document Creation (Customer Level) Until Arrival at the Depot That Fills the Order

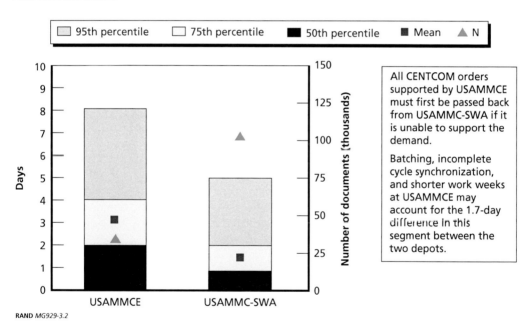

RAND MG929-3.2

---

[1] For a full discussion and treatment of this topic, see Addison et al. (2008)

[2] For USAMMCE FY2008, there were a total of 78,814 shipments to CENTCOM, of which 37,913 hit RFID and 28,765 had a valid destination "ping" where they were positively identified in the information system as arriving. For USAMMC-SWA FY2008, there were a total of 132,746 shipments to CENTCOM, of which 103,767 were in GATES and had a valid arrival date stamp. See Appendix E for a breakout of the population of data for each time-stamped, or identified segment, of the distribution process that was used to determine distribution performance.

middle (yellow) segment shows the 75th percentile time, and the height of the column shows the 95th percentile time, with the red square showing the mean time. There is close to a two-day difference in this segment for USCENTCOM customer orders filled by USAMMCE compared to those filled by USAMMC-SWA, with a significant difference in variability.

The sources of delay and variability are driven primarily by the two-echelon system and include:

- Requisitions may be held at USAMMC-SWA in anticipation of their being filled only to be passed back to USAMMCE if that proves not possible.[3]
- Passing documents from USAMMC-SWA to USAMMCE by batch process, instead of in real time, creates requisition hold time at USAMMC-SWA, and therefore adds to the total time before USAMMCE has the requisition. If requisitions are received at USAMMCE on weekends, considerable time will pass before the documents are pulled into USAMMCE's system.

The second process covers the time from requisition receipt to MRO. Producing an MRO involves a computer process that matches orders to inventory currently available at the depot. For this study, focusing on distribution processes, we attempted to limit cases to items immediately on hand at the depot (i.e., no backorders). This is typically indicated by an MRO type code of "B." While this works well for analyzing USAMMC-SWA issues where requisitions are never held in backorder status, it works less well at USAMMCE. For example, when a requisition there is partially filled from stock and the remainder is backordered; when stock from a vendor is received to fill the remainder of the order, the resulting second MRO is coded as "B" instead of "D" (backorder) if it is released via the routine "demand accommodation" process.

It is not possible to distinguish cases like this from "true" immediate issues. In our analysis, we include all MRO type "B" cases, acknowledging that this includes some number of backorders and falsely inflates the average time via these outliers, as Figure 3.3 shows. At USAMMCE, the median and 75th percentile times are both zero, while the 95th percentile is 20 days, driving the overall average to 2.5 days there versus less than one day at USAMMC-SWA, which actually has a median time of one day. This difference in the mean time is thus likely an artifact of measurement difficulties arising from miscoded immediate issues and not from a real difference in process performance.

---

[3] Discussions with military leadership at USAMMCE and USAMMC-SWA included mention of this as a practice. The requisitions are not held indefinitely, nor for a period of time detrimental to the customer, but instead when it is estimated that an anticipated replenishment and subsequent fill from USAMMC-SWA would be more responsive than a pass back and fill from USAMMCE.

Figure 3.3
Comparison of Times from Requisition Receipt to MRO

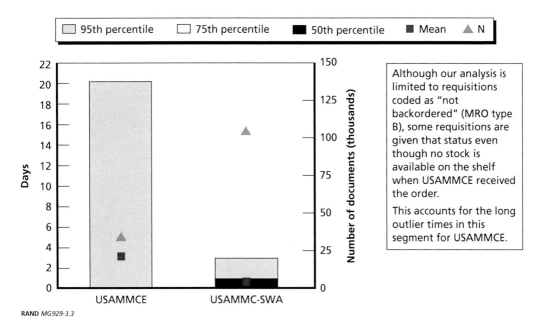

RAND *MG929-3.3*

Once MROs are cut, depot personnel retrieve items from their bins and shelves and send them to the consolidation locations. As shown in Figure 3.4, both USAMMC-SWA and USAMMCE are quite efficient at this part of the process, with USAMMC-SWA clearing its backlog of MROs every day, and USAMMCE filling the large majority of MROs for USCENTCOM the same day they are received, reflected in the 75th percentile time of zero. However, because USAMMCE does not run weekend shifts, some MROs wait until after a weekend or a holiday to get picked. A small number take eight days or more to get picked, as the chart above shows, although we cannot explain why the process would take that long for this small number of cases.

The most significant difference in times between distribution from USAMMCE and USAMMC-SWA comes from differences in load consolidation times, again driven by the two-echelon structure. In this segment, picked shipments are placed in customer-specific "triwalls" (large cardboard boxes capable of holding several hundred pounds of materiel). When sufficient volume is generated, the triwall is closed and sealed, with a radio frequency identification tag applied, and offered up for shipment via military or commercial air.

As seen in Figure 3.5, consolidation times are considerably longer at USAMMCE than at USAMMC-SWA, where most triwalls are sealed and offered for shipment the same day the items are picked.

Figure 3.4
Comparison of Times from MRO to Picking of Order

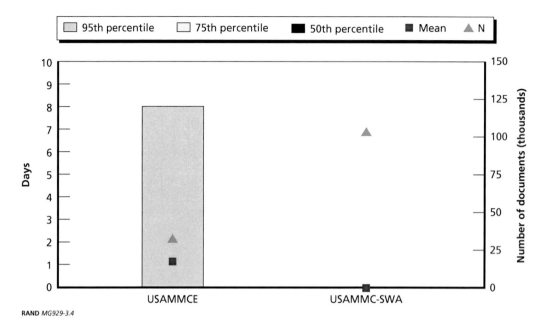

RAND *MG929-3.4*

Figure 3.5
Comparison of Times from Distribution from USAMMCE and USAMMC-SWA and Load Consolidation

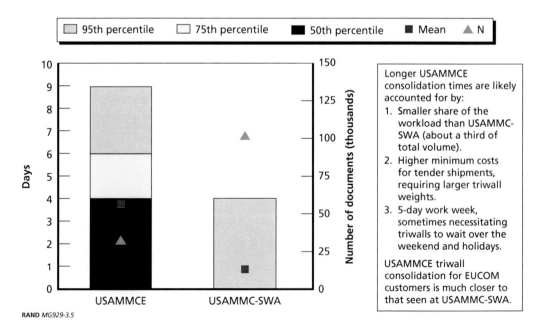

RAND *MG929-3.5*

There are several reasons for the relatively longer times at USAMMC:

- USAMMC-SWA fills about 60 percent of USCENTCOM orders and close to two-thirds by weight. Thus, USAMMCE accounts for much less of USCENT-COM volume by weight than USAMMC-SWA, leading to longer times to generate the same volume of weight and fill a box to a reasonable level.
- Minimum Class VIII tender airlift rates affect consolidation practices and times. For the period measured here (FY2008), all USAMMCE shipments were shipped via Class VIII tender while all USAMMC-SWA shipments were shipped via military air. The tender had minimum charges for shipments (ranging from $800 to $1,500), regardless of the shipment size and weight, resulting in a fixed, minimum charge for triwalls under 300 pounds or so. Air Mobility Command rates, by contrast, include no such minimum charge. To keep costs down, then, USAMMCE would have had to maximize volume in its triwalls or at least ensure they were relatively full.
- Shorter workweeks at USAMMCE extend consolidation hold time. Triwalls that do not have sufficient volume at the end of the workday are kept open while awaiting more items. This leads to longer consolidation times, especially when weekends or holidays intervene.

Thus, the longer delays at USAMMCE are primarily a function of supporting USCENTCOM customers via the Class VIII tender, and secondarily related to slower volume accumulation and shorter work weeks. As will be shown later, USAMMCE consolidation hold times for USEUCOM customers tend to be far shorter, because it is the first source for these customers.

In FY2008, aggregate transportation times from either depot to destinations in USCENTCOM were very similar, as shown in Figure 3.6. On average, the two were almost exactly the same. There were differences by location (discussed later with FY2009 data), with USAMMCE times longer to Afghanistan and USAMMC-SWA times longer to outlying locations in Iraq supported by military air.

As the preceding segment-by-segment analysis shows, USAMMCE times by segment tend to be longer than those at USAMMC-SWA, but these differences are mostly driven by the work week (five days/week compared to seven) and by the support structure design, which splits sourcing between the two. The end result is that USAMMCE times to USCENTCOM customers are significantly longer than those from USAMMC-SWA.

Figure 3.7 shows that difference: times to USCENTCOM, excluding the transportation segment, averaged a little over 11 days from USAMMCE and just over 3 days from USAMMC-SWA. There are large differences between the two in the time from requisition creation to depot arrival (driven by batching, pass backs, and USAMMCE's

Figure 3.6
Comparison of Aggregate Transit Times

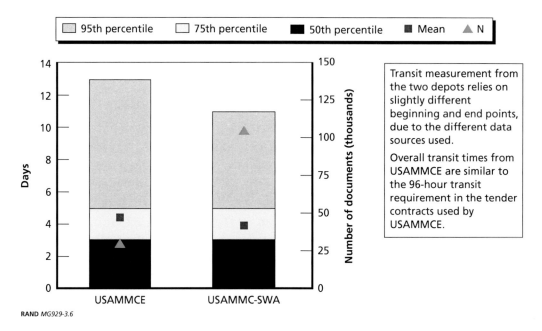

RAND *MG929-3.6*

Figure 3.7
Comparison of Distribution Processing Times

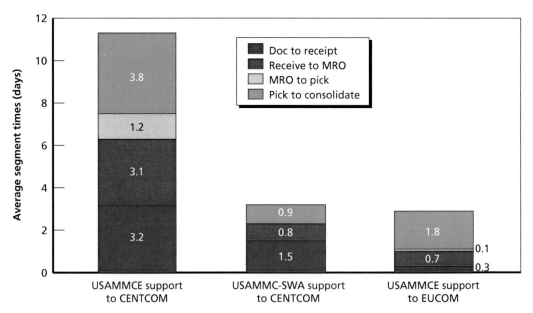

RAND *MG929-3.7*

schedule), time to cut the MRO (an artificial result of the difficulty in culling all back-orders from USAMMCE), pick/pack time (again influenced by USAMMCE's shorter work week), and most importantly by USAMMCE's longer hold time for consolidation (driven by cargo accumulation time from being the second source, the work week, and airlift rate structures).

As described, these differences are not inherent in USAMMCE's processes, however. They come mainly from USAMMCE's current secondary role in supporting USCENTCOM. When USAMMCE is the sole source of support for customers, performance looks rather different. As evidence of that, the column on the far right of Figure 3.7 shows FY2008 performance for USAMMCE in support of major USEUCOM customers. Overall times (again excluding transportation) are about the same as for USAMMC-SWA for its USCENTCOM customers.[4] Here we see no requisition pass-back delay. MRO release time is the same as at USAMMC-SWA.

The findings suggest that there are fewer split shipments for USEUCOM customers and highlights the fact that the requisition population is dominated by more easily filled high-demand items, whereas most of USAMMCE's issues to USCENTCOM customers are for lower-demand items, with USAMMC-SWA stocking the higher-demand items. Most important, with the Class VIII tender minimum size constraint not present, the USAMMCE consolidation time for USEUCOM customers is far less.

Figure 3.8 reinforces the points from the previous page; the lower mean times for USAMMCE shipments to USEUCOM customers than for USCENTCOM customers are associated with less process variability.

This analysis leads us to understand that split sourcing to fill USCENTCOM demand between the two depots has a significant effect on aggregate support to customers. As shown in Figure 3.9, USAMMC-SWA offers very good performance for the high-demand items stocked there. However, those items comprise no more than 60 percent of total USCENTCOM demand; the rest of customer requisitions are passed back to USAMMCE to be filled. That means that aggregate support performance for customers is a weighted average of the performance from both locations, as illustrated in the right-most column of Figure 3.9 (black border). USAMMC-SWA's average time for its 60 percent share is about 7 days, which combined with USAMMCE's 40 percent share averaging almost 16 days, yields an average time for USCENTCOM customers of about 10.5 days.

Supporting USCENTCOM customers predominantly or totally out of one location would likely yield significantly better performance by greatly reducing the delay induced by split sourcing.

The potential benefits of consolidating support at either location could be a 20 percent improvement in average end-to-end time, as illustrated in Figure 3.10. In this

---

[4] We have no means of measuring transportation times from USAMMCE to its USEUCOM customers, so we exclude showing the transportation segment in all three cases in the chart.

Figure 3.8
Comparison of Times for USAMMCE Shipments to USEUCOM Customers and USCENTCOM Customers

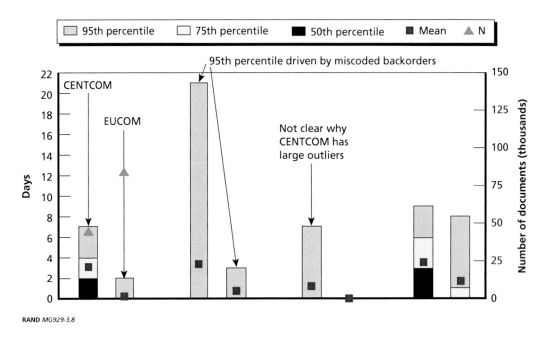

RAND *MG929-3.8*

Figure 3.9
Comparison of Average Aggregate End-to-End Times Over Ten Days to CENTCOM

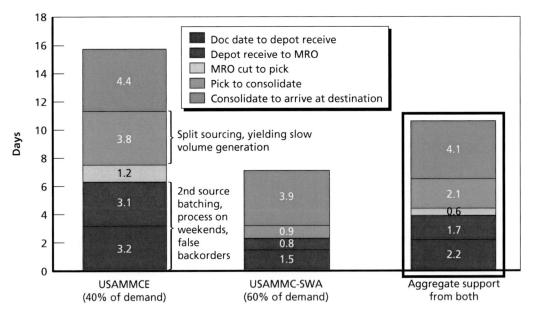

RAND *MG929-3.9*

Figure 3.10
Comparison of Aggregate Support to CENTCOM and One-Location Support

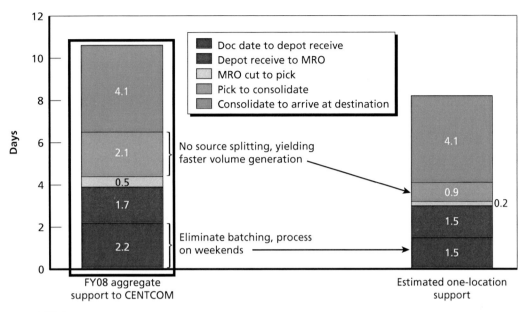

scenario, the pass-back delay would be eliminated. With a seven-day work week, we would expect to see fewer delays in cutting the MRO and picking the order. We would expect to see a significant improvement in consolidation time, as all demand would be funneled to a single source. Finally, we anticipate no degradation in transportation time, in the expectation of some greater efficiency in managing relations with the tender strictly from one location.

If USAMMCE were the sole source, it would be necessary to move to a seven-day/week schedule to support ongoing war operations as assumed in this figure. Whether this would be done, as at USAMMC-SWA, with deployed military personnel or by hiring local nationals would need to be resolved.

If support were consolidated at USAMMC-SWA, it would require a higher fill rate. Increasing the fill rate at USAMMC-SWA would require a much broader range of stocked items, significantly above the fewer than 3,000 items currently held at USAMMC-SWA. To increase USAMMC-SWA's fill rate from 60 percent to 85 percent would require about a tripling of the number of part numbers stocked.

If USAMMC-SWA's fill rate approached 85 percent, then USAMMCE's distribution strategy for direct support to USCENTCOM customers would have to change. It could no longer hold cargo for Class VIII tender triwalls on a routine basis; the consolidation times would become unsupportable. Instead, for its much smaller part of the workload, it would have to use alternatives, such as sending individual shipments loose to Ramstein Air Force Base to be lifted by Air Mobility Command or by using the World Wide Express contract for premium service for packages under 300 pounds.

Costs

This section estimates how costs would change were support to be consolidated at USAMMCE. We begin with a brief discussion of the FY2008 transportation model and approach for calculating cost.

The FY2008 Class VIII transportation model, shown in Figure 3.11, consisted of USAMMCE sending almost all of its shipments (98 percent of total weight) to the USCENTCOM AOR via Class VIII commercial tenders. USAMMC-SWA relied heavily on organic military airlift (supplemented by the Theater Express tender) along with a low percentage of total weight moving via Blanket Purchase Agreements (BPA) with commercial vendors. In August 2008, USTRANSCOM revised the Class VIII commercial tender contracts to include explicit rate quotes and weight breaks for shipments originating from USAMMC-SWA.

The lower part of Figure 3.12 depicts the FY2009 transportation model for USAMMCE and USAMMC-SWA. USAMMCE continued to use the Class VIII commercial tender to move a high percentage of the total weight the depot supplies to customers in USCENTCOM. However, USAMMC-SWA used the Class VIII tender for roughly 40 percent of its shipments, in terms of weight. Note that the Class VIII tender shipments move materiel from either depot to the same theater carrier

Figure 3.11
FY2008 Transportation Model

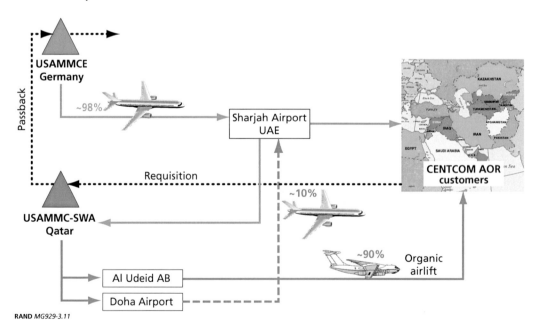

RAND MG929-3.11

Figure 3.12
FY2009 Transportation Model for USAMMCE and USAMMC-SWA and Its Effect on Costs

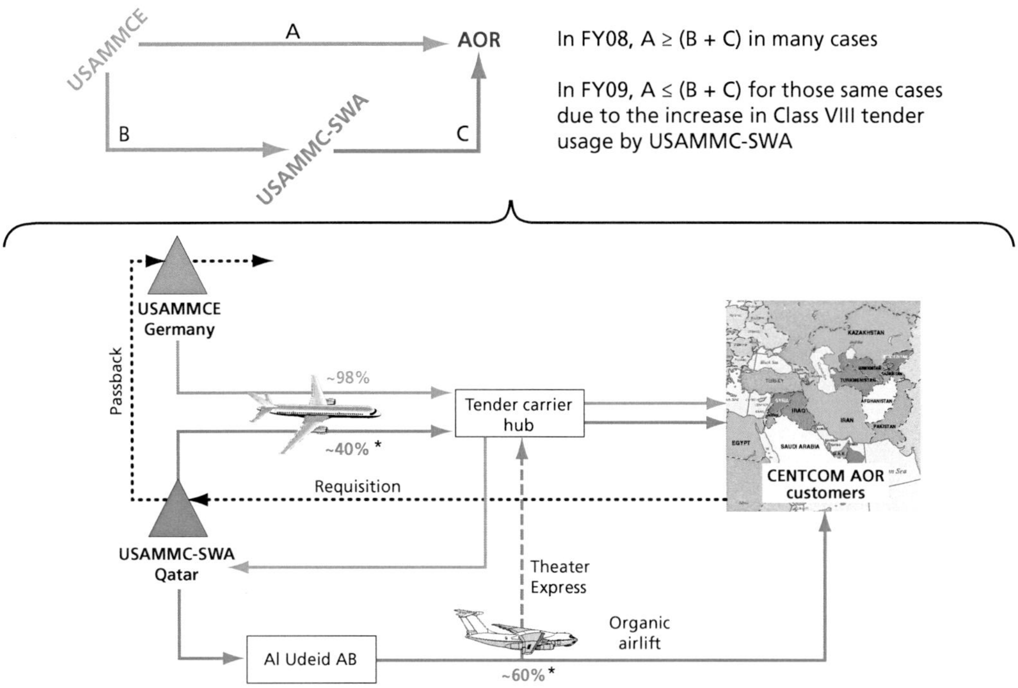

In FY08, A ≥ (B + C) in many cases

In FY09, A ≤ (B + C) for those same cases due to the increase in Class VIII tender usage by USAMMC-SWA

*USAMMC-SWA ratio of tender to total rose in January–February 2009 to ~55%.
RAND MG929-3.12

hub for final shipment to the destination airfield. Additionally, replenishments from USAMMCE to USAMMC-SWA are shipped through this same hub.

On the top portion of Figure 3.12 is a schematic to support a discussion of the effect this shift in airlift providers has had on the cost comparison between USAMMCE and USAMMC-SWA from FY2008 to FY2009, which changed as a direct result of the increased use of the Class VIII tender by USAMMC-SWA. Under the FY2008 transportation model, it was often more expensive to send items directly from USAMMCE to customers in the USCENTCOM AOR (Leg A), than it was to pay for the replenishment airlift leg from USAMMCE to USAMMC-SWA (Leg B) plus the distribution leg from USAMMC-SWA to the customer (Leg C). That is, in FY2008, it was common for the Cost of Leg A ≥ (Cost of Leg B + Cost of Leg C). With the introduction of USAMMC-SWA as a point of origin for the Class VIII tender in August 2008, some of the same cases where A ≥ B + C reversed and became cases in which A ≤ B + C: direct shipment from USAMMCE to the customer became less than the sum of the two-stage transportation involving USAMMC-SWA. The main driver for this reversal is the increased use of the Class VIII tender by USAMMC-SWA. Data received from USTRANSCOM indicate that the Class VIII tender was used for roughly 40 percent of the weight originating from USAMMC-SWA for the period October through December 2008. Military officers at USAMMC-SWA indicated that the usage of the tender increased to as much as 53 percent of total weight moved for the period January through February 2009.

In FY2008 and before, the relative "cheapness" of military airlift (AMC organic military aircraft and Theater Express) for USCENTCOM-bound shipments originating from USAMMC-SWA versus those originating at USAMMCE favored the two-leg transportation structure, with replenishment from USAMMCE to USAMMC-SWA and then shipment from USAMMC-SWA to the USCENTCOM customer. In late FY2008, the Class VIII commercial tender for airlift was rewritten to include rates for shipments to customers originating from USAMMC-SWA.

Figure 3.13 shows the amount charged against the Class VIII commercial tender contract for replenishment shipments from USAMMCE to USAMMC-SWA (blue columns) on a monthly basis. The monthly replenishment weights corresponding to these charges are shown by the red line with the scale on the right y-axis. As USAMMC-SWA began using the Class VIII tender, the associated charges and corresponding weights appear in the plot as a green column and a black line, respectively. Figure 3.13 serves as a visual depiction of the increase in double-handling of materiel by the commercial vendors. In other words, as USAMMC-SWA increases the use of the Class VIII

Figure 3.13
Cost of Class III Commercial Tender Contracts for Replenishment Shipments from USAMMCE to USAMMC-SWA

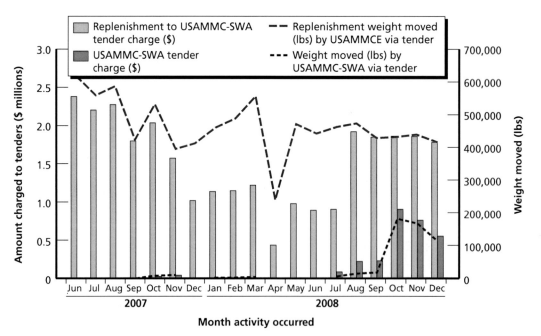

tender, the commercial vendors increasingly ship some of the same items twice through the same transfer hub. This increase in tender usage by USAMMC-SWA also eliminates the previously presented advantage of the two-leg transportation structure, because the Class VIII tender service is more expensive per pound moved than organic airlift.

In October 2008, USAMMC-SWA began making considerable use of the Class VIII tender. Table 3.1 shows the adoption of the tender through the changes in airlift mix over time. While the USAMMC-SWA ratio of Class VIII tender to total airlift leveled off at around 40 percent (October–December 2008), the team at USAMMC-SWA actively tuned the specific airlift mix by location during this period in an effort to improve performance and reduce cost.

Table 3.2 provides the FY2008 costs for the two Class VIII depots: USAMMCE and USAMMC-SWA. We extracted these data from the sources detailed earlier in the report.[5]

For USAMMCE, the cost of personnel represents the cost of the active-duty military ($2.18 million) and local nationals ($10.97 million) responsible for warehousing and distribution. The equipment and facilities maintenance costs were $150,000. There were no additional contract costs, nor additional facilities requirements at USAMMCE to include in this analysis. It is worth mentioning that there is discussion internal to the U.S. Army regarding the possible relocation of the USAMMCE operation. Our analysis assumes that this would not affect costs (or capability).

The number of pounds moved and number of issues represent the values used to measure the depot workloads. In FY2008, USAMMCE moved 9.60 million pounds through 49,000 issues and USAMMC-SWA moved 4.98 million pounds through 180,000 issues. It is important to recognize that the weight moved by USAMMC-SWA was first handled at USAMMCE. Therefore, the weight issued by USAMMCE

Table 3.1
USAMMC-SWA Airlift Mix, August to December 2008

Month	Commercial Tender	MilAir
August 2008	3.11 percent	96.89 percent
September 2008	4.67 percent	95.33 percent
October 2008	43.37 percent	56.63 percent
November 2008	42.14 percent	57.86 percent
December 2008	36.63 percent	63.37 percent

[5] For USAMMC-SWA, the cost of personnel represents the cost of the active-duty military personnel ($2.57 million) and contractors (Eagle Contract at $2.66 million) responsible for warehousing and distribution. The cost of equipment and facilities we received from USAMMC-SWA was $230,000, which covers maintenance and operating costs of equipment at the depot. There were no additional contract costs, nor additional facilities requirements at USAMMC-SWA to include in this analysis.

Table 3.2
Cost Actuals for Personnel, Facilities, and Equipment Used as Comparison Baseline

	USAMMCE	USAMMC-SWA
Class of materiel	VIII	VIII
Cost of personnel (FY2008)	$13.14 million	$5.23 million
Cost of equipment (FY2008)	$0.15 million	$0.23 million
Additional contract costs (FY2008)[a]		
New facilities cost [b]		
Number of pounds moved (FY2008)[c]	9,601,494	4,981,018
Number of issues (FY2008)[d]	485,282	180,451

[a] No additional contract costs were incurred at either depot in FY2008; these costs could include employee overtime and other costs which are not already captured in the contract costs included in the cost of personnel.

[b] No additional construction is necessary at either location to perform the USAMMC-SWA mission.

[c] Total weight (pounds) moved indeterminate of delivery shipping method (ground or air).

[d] Total number of issues completed at a depot indeterminate for delivery shipping method (ground or air).

would not change were it to become the sole source of direct support to USCENT-COM customers. Issues would go up some as higher-quantity replenishments would be replaced by lower-quantity customer issues. We use the USAMMC-SWA customer issue and replenishment volumes to estimate the change in USAMMCE issue work-load. The weight moved by USAMMC-SWA was also decomposed by location by month so as to accurately estimate the cost of transportation if the items were to have been supplied directly from USAMMCE.

To estimate how USAMMCE costs would change with the increase in issues, we estimated the increase in issues and multiplied it by the cost per issue, as shown in the left side of Figure 3.14. We performed this calculation assuming the adoption of a seven-day work week at USAMMCE. The USAMMC-SWA variable in the equation at the bottom of the figure is populated using the actual USAMMC-SWA cost, which is dominated by the cost of personnel.

To calculate the "transportation delta," the weight moved by USAMMC-SWA was decomposed by country (Iraq and Afghanistan) by month. We then applied the USAMMCE transportation rates to these same countries. Since USAMMCE is already an integrated part of the Class VIII supply chain, it is not necessary to proj-ect the USAMMC-SWA airlift mix onto USAMMCE. Instead, we assumed that USAMMCE would continue to exclusively use the Class VIII tender to serve all cus-tomers. Note that for the transportation rate portion of the analysis we included the average rate associated with inventory replenishment (across October–December 2008) in the USAMMC-SWA plotted rates.

Figure 3.14
Comparable Cost Analysis Methodology: USAMMC-SWA versus USAMMCE

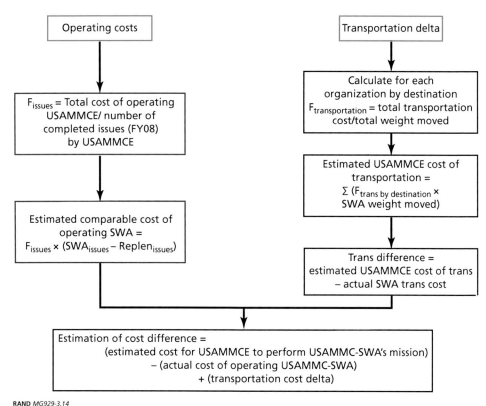

RAND *MG929-3.14*

Figure 3.15 shows the costs of performing the USAMMC-SWA mission from USAMMCE. This chart is similar to Figure 2.7, which was used for the DDKS to USAMMC-SWA cost comparison. At the bottom of the figure is the cost comparison equation: the estimated cost for USAMMCE to execute SWA's mission minus the actual operating costs of SWA plus the change in transportation costs.

On the left side of the left graph, we show USAMMC-SWA's operating costs per month. In the same graph, we show the estimated monthly operating cost increase at USAMMCE were it to perform USAMMC-SWA's mission.

On the right side of the chart is a graph comparing the transportation rates. The columns represent the aggregate transportation rates to depot customers in Afghanistan and Iraq/Kuwait via military airlift, military contract airlift (Theater Express), and Class VIII tender airlift from October through December 2008.[6]

---

[6] Forty percent of the airlift out of USAMMC-SWA is via the Class VIII tender, while the remaining 60 percent consists of a two-thirds, one-third mix between Theater Express and organic military airlift, respectively. This mix was determined using data for the Theater Express (supplied by USTRANS-

Figure 3.15
Costs of Performing USAMMC-SWA Mission from USAMMCE

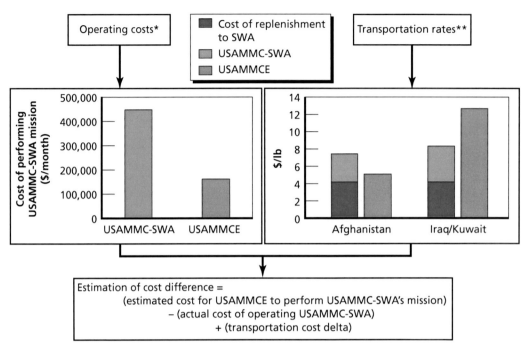

*Monthly operating costs determined using FY08 annual data and transaction costs at each location with replenishment to SWA replaced by USAMMCE issues.

**Transportation rates using October–December 2008 data; rates are not at steady state as USAMMC-SWA ratio of tender to total continues to increase.

RAND *MG929-3.15*

The dark blue lower portions of the left columns indicate the cost of replenishment to SWA. The light blue upper portions of the left columns provide the cost of going from SWA to the customer. Transportation for USAMMCE to Afghanistan costs less than from USAMMC-SWA (including replenishment airlift) and vice versa for Kuwait and Iraq.

Figure 3.16 shows the estimate of how costs would change were support to be consolidated at USAMMCE, with the top bar showing data for October–December 2008, and other bars showing the resulting estimate with changes in the demand mix between Iraq and Afghanistan. Both the USAMMC-SWA monthly operating cost and the estimated cost to perform the USAMMC-SWA mission on a monthly basis from USAMMCE are constants in the cost analysis.

COM) and the organic military airlift weight percentage for USCENTCOM provided by the Tanker Airlift Control Center (TACC) within the Air Force's AMC.

Figure 3.16
Transportation Costs with (A) Decrease in Shipments to Iraq, (B) Increase in Shipments to Afghanistan

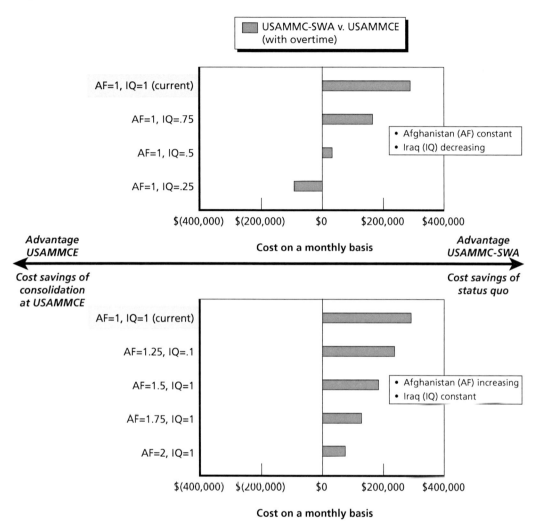

NOTE: Cost of replenishment is included for SWA for all cases.
RAND MG929-3.16

The "transportation delta" variable in the cost advantage equation is what drives the movement in the cost comparison. Variation is caused by the transportation rates (to Iraq and to Afghanistan) from each depot to each destination applied to the weight moved out of USAMMC-SWA to these locations. A bar displaying value to the right of the center vertical axis indicates that there is a cost advantage to continuing to supply USCENTCOM Class VIII materiel from USAMMC-SWA. Conversely, a bar pointing to the left indicates that performing the entire USCENTCOM Class VIII mission from USAMMCE would generate a cost savings over continuing the mission from USAMMC-SWA.

The two plots in Figure 3.16 show the independent effect of the transportation rates to Iraq and to Afghanistan on the overall cost advantage determination. The baseline case (early FY2009) is based upon the average monthly Class VIII weights moved using October to December 2008. The top plot shows that a decrease in weight shipped to Iraq, while holding weights to Afghanistan constant, drives the cost advantage toward consolidation at USAMMCE. Increasing Afghanistan, while holding Iraq constant, has less of an impact on the cost comparison, as shown in the lower plot. There is a greater effect from altering the weight shipped to Iraq versus changing the weight shipped to Afghanistan.

Figure 3.17 shows the effect of both increasing shipments to Afghanistan and decreasing them to Iraq. With an increase in troop levels in Afghanistan to 1.5 times the number in late 2008, and Iraq levels decreasing in 25-percent steps, the cost comparison shifts from favoring maintenance of the status quo to favoring consolidation at USAMMCE. When the number of troops in Afghanistan is doubled from the early FY2009 levels, and troop levels in Iraq are at one quarter of early FY2009, the cost comparison indicates an even stronger case for consolidation at USAMMCE, with savings of about $300,000 per month.

Figure 3.17
Combined Effect on Costs

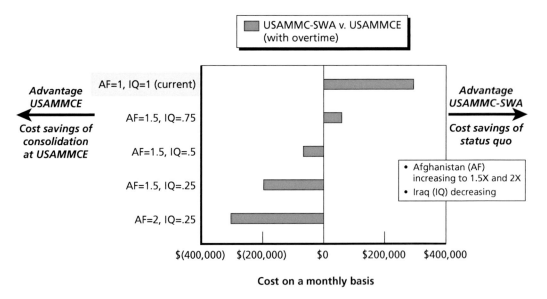

NOTE: Cost of replenishment is included for SWA for all cases.
RAND MG929-3.17

A key factor in this shift is that the decrease in troop levels in Iraq would result in a corresponding decrease in the use of National Air Cargo (NAC), which is the preferred Class VIII tender from USAMMCE to Iraq. NAC's preferred status is directly related to the company's ability to provide unmatched performance. But this improved performance carries with it a cost premium, which is the factor directly responsible for the cost advantage favoring USAMMC-SWA in the "current" case. As the extreme case is reached (Afghanistan troop levels equal two times "current" and Iraq levels equal one-quarter of "current"), a $300,000-per-month cost advantage for consolidation at USAMMCE develops.

Through the course of the project, we were presented with background information on the establishment of USAMMC-SWA as a provisional organization to support the USCENTCOM AOR along with the associated discussion and planning for the ultimate closure of the depot. One metric that could be used to measure the cost-effectiveness of maintaining USAMMC-SWA is the cost per issue out of the depot. Appendix H offers a detailed analysis of cost per issue as it develops over time from USAMMC-SWA within a changing USCENTCOM AOR.

Transportation performance from USAMMCE to Iraq is similar to transportation performance from USAMMC-SWA but more expensive. This is directly driven by the use of NAC, which gives better performance from USAMMCE to Iraq than other carriers, which is the reason for the choice of carriers. USAMMCE transportation performance to Afghanistan is slower, but it is also less expensive.

Table 3.3 shows that the use of a different carrier can affect cost, and also performance. The left black box shows that using NAC from USAMMCE costs $8.45 per pound, or $4.67 more than UPS, while providing a performance advantage of 2.5 days (third black box).

As is evident in the data, and validated through interviews, USAMMC-SWA appeared to be actively managing its use of the Class VIII tender and was able to improve cost-effectiveness from October to December 2008. As the use of the tender

Table 3.3
Costs per Pound and Average Transportation Times to Iraq and Afghanistan Using Different Carriers

From USAMMCE	Cost per Pound				Average Transportation Time (days)			
	To Iraq		To Afghanistan		To Iraq		To Afghanistan	
		NAC Disadvantage ($/lbs)		NAC Disadvantage ($/lbs)		NAC Advantage (days)		NAC Advantage (days)
NAC	$8.45		$4.75		3.2		3.8	
UPS	$3.78	$(4.67)	$3.86	$(0.89)	5.7	2.5	5.2	1.4
DHL	—	—	$3.28	$(0.58)	—	—	6.2	2.4

increased, the cost initially increased. Over the course of three months, USAMMC-SWA leadership maintained the ratio of tender to total airlift but also decreased costs by fine-tuning the mix of airlift providers by location.

Other Considerations for Consolidating

There are other capabilities to consider when contemplating the closure of USAMMC-SWA and consolidation at USAMMCE. Besides materiel warehousing and distribution, there are medical maintenance and repair activities, forward repair activity mission (FRA-M) teams that use USAMMC-SWA as a base of operations, patient movement item (PMI) cell support, optical fabrication, and customer support operations to consider.

While there is very little empirical, objective data available to ascertain the impact of relocating these activities to USAMMCE, we qualitatively assess the impacts. Additionally, we employed perceptions from the field that were collected in interviews with USAMMC-SWA customers (medical logisticians based forward in the USCENTCOM AOR) during a medical logistics conference at Camp As Sayliyah in October 2008. For medical maintenance we used the available data within the Integrated Logistics Analysis Program (ILAP) to help develop an understanding of the potential impact of closing maintenance capabilities at USAMMC-SWA.

To assess the impact of closing or duplicating maintenance facilities and capabilities, we examined workload data for the repair of medical equipment at several locations in Southwest Asia and USAMMCE. The medical maintenance and equipment repair data were extracted from the ILAP for the 30th MedLog.[7]

For the purposes of our analysis, we focused on the Maintenance UIC field, which records the location where the maintenance was performed, and the Unit UIC field, which records the unit that sent the item to the maintenance location.[8] We then examined the data in three primary ways, as shown in Figure 3.18. We first evaluated the number of repairs performed at each of the medical maintenance locations. USAMMCE performed 52 percent of all maintenance actions, while USAMMC-SWA performed approximately 31 percent of all maintenance actions. All of the remaining maintenance locations individually contributed less than 10 percent of the repairs. This

---

[7]   A total of 7,884 maintenance records were gathered for maintenance actions that were completed over the time horizon of January 2008 through mid-February 2009. The majority of these records are from USCENTCOM customers. Less than 30 of the maintenance actions began in 2007, and the preponderance of the data are for medical repair actions that took place in 2008.

[8]   We determined 7,568 records to be unique maintenance events. We define a unique maintenance event as one in which a unit sends an item to a maintenance location, and the location repairs the item. In addition, we identified 316 records that show double movements of specific items, meaning they were first sent to a maintenance location by an owning unit, and then were evacuated to another maintenance location for further inspection and repair. We therefore consider 316 movements of these items, but only 158 repairs, leaving the total data set of equipment repairs to be 7,726 in the data set.

Figure 3.18
Source of Equipment Repair by Number of Equipment Repairs

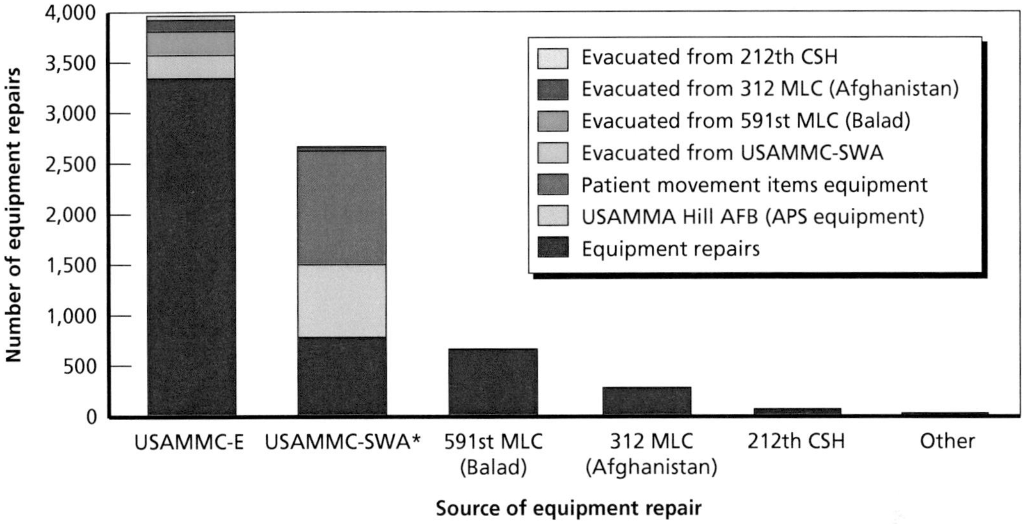

* Some USAMMC-SWA maintenance transactions may not appear in the ILAP data due to problems in the transition to SAMMS-2, therefore the number of transactions that appear here should not be assumed to be actual. However, the relationship in the number of transactions is the same—PMI is the largest source of maintenance action, with APS and equipment repairs making up the remainder.

SOURCE: ILAP, October 2007–February 2009.

RAND MG929-3.18

initial assessment revealed that the majority of maintenance actions for USCENT-COM customers are being performed at USAMMCE and USAMMC-SWA.

We then examined how many of the repairs were evacuated to each maintenance location from *other* medical maintenance locations. We considered all repairs that were generated directly from maintenance locations, and the 158 repairs that were sent to a maintenance location and were then evacuated to USAMMCE. Over 600 repairs, or 16 percent of the total repairs at USAMMCE, were evacuations from other locations, likely due to more comprehensive repair capabilities.

Finally, we explored the types of items being repaired at USAMMC-SWA. We found that approximately 70 percent of all the maintenance actions performed at SWA are for Patient Movement Items (PMI) and Army Prepositioned Stock (APS) items: 43 percent and 27 percent, respectively. Because the maintenance and repair of these items could be performed at any location, a decision to close USAMMC-SWA and in conjunction reduce or eliminate capabilities of USAMMC-SWA medical maintenance repair capabilities would be feasible, requiring a decision of where to perform the PMI and APS missions that currently reside at USAMMC-SWA.

Table 3.4 provides the qualitative assessments of closing operations at USAMMC-SWA. Based upon the data in Figure 3.18, we do not foresee any negative impact on

Table 3.4
Qualitative Assessment of the Effects of Closing Operations at USAMMC-SWA

SWA Capabilities	Option	Implications		
		Performance	Cost	Intangibles
Medical equipment maintenance and repair	Move to USAMMCE	No known impact	May reduce cost of repair part inventory	May increase cross-training; will have access to ISO 9000 facilities
FRA-M mission support	Move to USAMMCE or Balad	No known impact	No known impact	The FRA-M team only needs a bed-down location
Patient movement item (PMI) cell support	Move to point of sortie origin or destination (e.g., Ramstein)	No known impact	No known impact	
Optical fabrication	Move to USAMMCE	No known impact	No known impact	
Customer and contingency operations support	Move to USAMMCE	No known impact	No known impact	May not have support that is fully "attuned" to theater environment

performance—that is, service to customers in the USCENTCOM AOR—if medical maintenance were consolidated at USAMMCE. In contrast, there might be potential benefits. The repair facilities at USAMMCE are ISO 9000 certified, and centralizing repair parts inventory at one location could reduce the overall cost of this inventory. Centralizing repair technicians could also facilitate cross-training among technicians and provide more time on equipment for repair experience.

When a piece of medical equipment fails, it is replaced by a like item that is kept in the operational readiness float (ORF). The purpose of the ORF is to get a replacement for a piece of equipment back into the field quickly so that readiness is not adversely impacted. Items that do not have an ORF inventory, or that are very large (such as a computed tomography—or CT—scanner) and cannot easily be evacuated to a repair facility, are tended to by forward repair activity mission (FRA-M) teams.

With respect to the forward repair mission teams, we did not find any data that would indicate performance degradation or an increase in costs if the FRA-M teams were not located at USAMMC-SWA. The FRA-M teams primarily need a base of operations for communication and to bed down; this could be at a number of places either within the USCENTCOM AOR, such as Camp Arifjan in Kuwait, or even at USAMMCE.

The Air Force PMI at USAMMC-SWA and the Army medical repair technicians provide diagnostic and service support. We found no data that could support a conclusion that moving PMI support from USAMMC-SWA to a different location would negatively affect performance or increase the cost of the operation. PMIs could be serviced at either the point of sortie origin or sortie destination such as Ramstein Air Force Base, where there is a high volume of patients moving out of the USCENTCOM AOR and which is in close proximity to USAMMCE.

Although optical fabrication is a convenience for soldiers who are at Camp As Sayliyah on rest and recreation leave, we did not find any evidence to suggest performance of the mission to provide eyeglasses or optical inserts would be negatively impacted were those operations moved out of USAMMC-SWA. Medical companies have optical fabrication capability, and there are several optical teams deployed in Iraq and Afghanistan.

Based upon our discussions during site visits in USCENTCOM, we note that there is a perception that if customer service is in the theater, the personnel providing support would be more "attuned" to the environmental conditions and what is happening in combat operations and therefore able to provide better service. However, we found no data or evidence that customer support operations would be negatively impacted or would cost more if they were moved out of USAMMC-SWA.

Summary of Consolidation at USAMMCE

Consolidation of USCENTCOM medical logistics at USAMMCE would result in lower costs with similar performance assuming the current shift in operations to Afghanistan and the future reduction in forces in Iraq. The performance of the non-transportation distribution process would improve with the elimination of the fragmentation of the supply network where sourcing is split and orders are passed back from USAMMC-SWA to USAMMCE.

The current transportation performance to Iraq from USAMMCE is similar to that of USAMMC-SWA; however, it is considerably more costly to ship to Iraq when NAC is used as a carrier. The current transportation performance from USAMMCE to Afghanistan, using UPS and DHL as carriers, is slower than from USAMMC-SWA. However, part of the performance gap between USAMMCE and USAMMC-SWA to Afghanistan could be the result of lower volumes of materiel being processed at USAMMCE. During our site visit to USAMMCE, we observed large boxes of materiel destined for Afghanistan being consolidated and waiting to be filled in order to justify the fixed cost associated with a shipment.

The benefits of the elimination of system fragmentation would outweigh increased transportation times to Afghanistan, leading to similar or better performance, with potential opportunity for improvement if transportation from USAMMCE to Afghanistan were improved through higher volume or increased management emphasis.

Cost of Replication at USAMMC-SWA

"Replication" of USAMMCE capabilities at USAMMC-SWA would improve performance by eliminating distribution network fragmentation, but there would be a cost penalty in comparison to consolidation at USAMMCE. To achieve this option, inventory investments would have to be made in order to reduce the pass-backs to USAMMCE. Currently, USAMMCE stocks approximately 13,000 lines of materiel, while USAMMC-SWA stocks approximately 3,000 of the fastest-moving lines. Based upon a rough estimate, an 85 percent customer service fill rate target would require approximately 5,600 additional lines (for a total of 8,600 lines to be stocked at USAMMC-SWA) at a total cost that would likely be less than $1 million.[9] Although there may not be enough space to accommodate the additional stock levels at USAMMC-SWA at its current location at Camp As Sayliyah, there is a request in to add an additional 30,000 square feet of space when operations are moved to Al Udeid Air Base by the fourth quarter of 2012. One potential complication is that if additional inventory were added to USAMMC-SWA and a customer service target of 85 percent fill was accomplished, there would still be 15 percent that would have to be satisfied by USAMMCE. This low volume of materiel might be a problem for the Class VIII tenders: it might not be enough for the service or to get the prices that are in effect at this time, and alternatives such as World Wide Express might have to be explored.

Along with more inventory, there would be a need for more personnel to handle and manage this inventory. First, additional warehouse personnel would be required to accommodate the stocking, picking, and packing of the added inventory. These particular personnel would likely be hired using the Eagle Contract with an estimated additional cost of $1.2 million per year for accommodation of 85 percent of demand, or $1.8 million for roughly 100 percent of demand.

Second, personnel would be needed to manage the medical air bridge supplying materiel coming out of CONUS, assuming direct replenishment as opposed to replenishment from USAMMCE stocks, as well as personnel to manage the new item requests (NIRs), which number in the hundreds per month at USAMMCE. As an alternative to locating the vendor support and NIR processing forward, it has been suggested within the medical logistics community that a capability within CONUS be established to remotely perform these activities.

---

[9] We derived the estimate for additional inventory cost associated with the replication at USAMMC-SWA option by looking at average demand for the additional lines of materiel and then pricing the cost of accommodating this demand by utilizing the respective unit prices for each line.

CHAPTER FOUR

Comparison of Options and Recommendations

There are three options that would preserve or improve performance and/or costs. The status quo would not change cost or performance. Next, operations could be consolidated at USAMMCE, resulting in better performance than the status quo and some reduction in cost. Finally, replication of USAMMCE at USAMMC-SWA would result in better performance than status quo, but there would be a slight increase in costs.[1]

Because distribution performance would be worse and costs would not be lower than USAMMC-SWA, the DDKS option does not meet the criteria for an option to be considered.[2] Additionally, because DDKS does not currently support medical materiel distribution, there would be a need for infrastructure investment such as cold chain assets and the integration of TAMMIS Enterprise Wide Logistics System (TEWLS), the medical logistics information system, neither of which we specifically accounted for due to the unattractiveness of the option even without considering these costs. Support from CONUS through a CCP would result in unacceptable performance and is therefore not an option that would meet the criteria for consideration.

We now compare the options that meet the performance and cost methodology requirements and discuss associated recommendations. Table 4.1 summarizes the transition requirements for the different options that meet the performance criterion.

The transition requirements vary considerably among the different options we consider for providing medical materiel to USCENTCOM customers. Obviously, maintaining the status quo requires no transition. Consolidating operations at USAMMCE would require going to a six- or seven-day per week operation. Somewhat replicating USAMMCE capabilities at USAMMCE-SWA would require establishing the medical air bridge capability to USAMMC-SWA, an increase in inventory, and some reconfiguration of TEWLS as USAMMC-SWA is currently indentured to USAMMCE for inventory replenishment transactions.

---

[1] In particular, the costs associated with transporting materiel to Afghanistan would be higher at a time when operations tempos in Afghanistan are increasing.

[2] There is some risk that costs could increase during the renegotiation of the DDKS contract in 2010.

Table 4.1
Transition Requirements by Option

Status quo: Requires no transition activities
Moving to USAMMCE: Require weekend shifts

Replication of USAMMCE capabilities at USAMMC-SWA requires:
• Establishment of medical air bridge from CONUS
    — Personnel to manage the prime vendor air bridge relationship, transactions and contracts
    — Personnel to manage new item requests
    — Virtual CONUS solution may mitigate the need for this support
• Increase in inventory
• Reconfigure TEWLS to make USAMMC-SWA a master plant
    — SAP programming, intermediate document interface exchange, expeditors & the establishment of a support office to manage the site.

Table 4.2 provides a comparison of all the options considered.

Table 4.2
Comparison of Options Considered

Option	Performance	Cost	Other Factors
Status quo	—	—	—
Consolidate at USAMMCE	Slightly better performance than status quo with elimination of pass-back delays and consolidation	Better cost efficiency	—
Replicate at USAMMC-SWA	Better performance than status quo	Potentially higher cost	Would need to establish and manage prime vendor support and new item request management[a]
DDKS	Worse performance to Afghanistan and Iraq	Likely similar cost—some risk of higher cost	Transition would create need for medical logistics–specific assets and medical logistics information system
CONUS Support – DDSP	Overall worse performance		Transition would create need for medical logistics–specific assets and medical logistics information system

NOTE: Shaded areas do not meet acceptability criteria.

[a] Establishing a CONUS capability to provide prime vendor and NIR support for deployed units might mitigate this personnel and management requirement.

A subjective risk assessment suggests that the three qualifying options have similarly low risk with respect to maintaining performance.

Status Quo. Maintaining the status quo does not introduce any new risks.

Consolidation at USAMMCE. If USAMMC-SWA was closed and operations consolidated at USAMMCE, the partial inventory buffer that currently exists to hedge against the adverse impact of disruptions in the supply chain would be removed. Right now, there are two sources of inventory for fast-moving items.[3] If USAMMC-SWA were to close, disruptions from USAMMCE would have greater impact. However, there are still alternative options in most cases. There are multiple transportation routes to most locations. Additionally, items could still be provided through prime vendors from CONUS if something were to happen to USAMMCE itself.

Replication of USAMMCE capabilities as USAMMC-SWA. This option would require management of the prime vendor medical air bridge as well as NIRs, increasing the training requirement and management complexity at USAMMC-SWA. This could present some risk with its dynamic personnel profile driven by the twelve-month deployments of military personnel and the handoff between deployments. However, this could be mitigated if a virtual, remote management capability residing in CONUS could be developed. There may also be some risk associated with the transportation network from USAMMCE if operations were consolidated at USAMMC-SWA. If the volume from USAMMCE is reduced, there is some risk of the transportation performance decreasing as a result of lower volumes, and alternatives such as World Wide Express may need to be explored. But this would have to be considered in light of improved performance for the items added to inventory at USAMMC-SWA.

Overall Conclusion

Consolidation at one location would yield about 20 percent better performance, and if at USAMMCE would likely provide for a relatively modest reduction in total costs, anywhere from $1.0 million to $3.5 million per year, depending upon the level of operations in Iraq and Afghanistan. Such consolidation could potentially further reduce costs and improve performance through renegotiation of the Class VIII tender contracts to provide all materiel distribution out of one airfield. Consolidation at USAMMC-SWA would improve performance, perhaps even more, but it would be more costly.

Army G-4, USAMRMC, and the Defense Medical Logistics Supply Chain Council all positively received the study results, acknowledged that medical logistics requirements are changing along with the troop levels in Iraq and Afghanistan, and requested that this study be updated in 12–18 months.

---

[3] The risk of disrupting the service for slower-moving items would not change.

Medical Logistics as a Distinct Discipline

Past studies have typically concluded that Class VIII (medical) supply is sufficiently unique and different from other supply classes to call for separate handling, distribution, and management. In particular, demand for the materiel is often urgent, there are legal mandates that govern their storage and control (such as Class IV narcotics), and some of the products are vulnerable to temperature changes, exposure to the elements, or degradation over time.[1]

The Hoover Commission Report of 1955 recognized that medical materiel is a highly specialized category of supply and must be procured, stored, and distributed differently from other types of materiel. In 1965, the Department of the Army Board of Inquiry on the Army Logistics System (Brown Board) recommended the formation of the Health Services Command to control all resources necessary for medical logistics, gave Army Materiel Command responsibility for Logistics except for medical, construction, and transportation, and recognized the need for logistics personnel forward in medical units due to the requirement for responsiveness (Ursone, 1988). A 1965–1969 Logistics Review of the U.S. Army written at the direction of LTG Mildren, Deputy Commanding General, U.S. Army, Vietnam indicated that:

1. The medical commodity has unique characteristics and requirements not easily blended into the general supply system;
2. The general supply system provided inadequate support during the consolidation period for Vietnam, with zero balances rising to 28 percent, and;
3. Experience gained in Vietnam proved that the medial commodity must remain under the control of the Surgeon to satisfy the needs of the physician (Memo from the Chief #23, "Medical Logistics" 19 April 1985).

---

[1] The studies are the "1953 Munitions Board Study of the Medical Supply System," the "1955 Hoover Commission Report," the "1965 Department of the Army Board of Inquiry on the Army Logistics System (Brown Board)," the "1965–1969 Logistics Review—U.S. Army Vietnam at the direction of LTG Mildren, Deputy Commanding General, U.S. Army Vietnam," the "1973 Bureau of Medicine and Surgery Study and Technical Workshop on Medical and Dental Supply Support," the "1985 Comptroller of the Army Installation Study," and the "1994 Department of the Army, DCSLOG Directed Analysis by the U.S. Army Logistics Evaluation Agency on Medical Logistics Policy Proponency."

Additionally, medical logistics personnel are often required to have product expertise such as special handling and control restriction training (in the case of "cold chain" and controlled narcotics), and they use a separate enterprise data system, TEWLS, that manages not only federal supply National Stock Numbers (NSNs) but also part numbers or item numbers that are also used in the commercial sector.

Payment of Commercial Tender Air Bills Comes Out of Medical Logistics Budget

Discussions with the personnel at USAMMCE, USAMMC-SWA, USTRANSCOM, and USCENTCOM yielded a flow chart (Figure B.1) that details how the Class VIII commercial tender vendors are invoiced and paid. Both USAMMCE and USAMMC-SWA follow the same processes and ultimately submit the Government Bills of Lading (GBLs) to the Defense Finance and Accounting Service (DFAS) office in Kaiserslautern, Germany. Currently, all Class VIII tender invoices (both intertheater and intratheater) are satisfied by funds supplied through Army G-4 Transportation Account Codes (TAC) supporting Operation Enduring Freedom (OEF) and Operation Iraqi Freedom (OIF). In June 2009, payments under the Class VIII tender were scheduled to

Figure B.1
Payment of Commercial Tender Air Bills

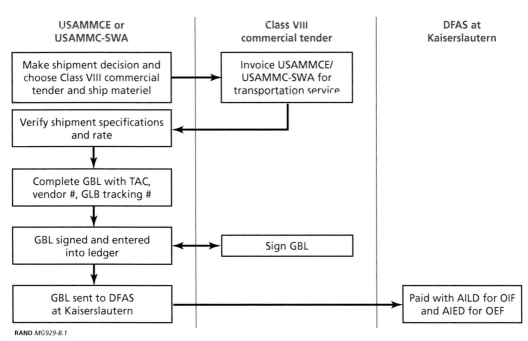

RAND *MG929-B.1*

begin to be processed using the PowerTrak automated invoice and payment system. As the use of the tender increases for intratheater airlift, it may become necessary for USCENTCOM to assume the responsibility for paying this portion of the tender charges.

APPENDIX C

Data Sources Used for Analysis

The data sources used for the analysis of distribution performance are shown in Figure C.1.

Figure C.1
Data Sources Used for Analysis

Moment in time	Assigned variable	USAMMC-E	USAMMC-SWA	Data source
Requisition created	"doc"	√	√	TAMMIS
Requisition received at depot	"rec"	√	√	TAMMIS
MRO printed for action	"iss"	√	√	TAMMIS
RFID tag burned	"wrdt"	√	√	RFID
Arrival of shipment into GATES	"POE"		√	GATES
Transit elapsed time (from wrdt or poe to arrival at destination APOD)	"transit"	√	√	GATES (SWA)* RFID (E)

Key, time-stamped event used as breakpoints to highlight data trends and phenomena.
*GATES (SWA) only for AMC-lifted shipments (including AMC tenders).

RAND *MG929-C.1*

Our Analysis Focused on Twenty Major Customers at Ten Destinations in Iraq, Afghanistan, and Kuwait

We limited our study of FY2008 distribution performance to the largest twenty USCENTCOM customers by demand. (See Figure D.1.) This is only a fraction of the "ship-to" addresses (roughly 32 percent) served by both USAMMC-SWA and USAMMCE; however, this small number accounted for the great majority of all customer orders to the two depots. In FY2008, for example, USAMMC-SWA filled orders for 411 customers and USAMMCE for 836. Most of these customers were small. For instance, 301 of USAMMC-SWA's customers had fewer than 100 total shipments, or 73 percent of USAMMC-SWA customers. Similarly, 544 of USAMMCE's customers had fewer than 100 total shipments, or 65 percent of USAMMCE customers.

Figure D.1
Our Analysis Focused on Twenty Major Customers at Ten Destinations in Iraq, Afghanistan, and Kuwait

Destination	Major ship-to addresses				USAMMC-SWA shipments	USAMMC shipments
Al Asad	W91532*	W91HYE**			3,578	1,496
Baghdad	FM6942	W91HYH**	W91Ka9	W91WY3	16,605	5,475
Bagram	FM6924	W91DCK			20,675	5,471
Balad	FM6943	W91947			34,200	10,045
BUCCA	W91JG7				3,954	1,216
Kirkuk	FM6938	W91PNV			4,212	794
Kuwait	FM6913	N68685	W81P37		9,708	6,376
TQ	M94207				7,561	1,151
Tallil	W91HYF				127	32
Tikrit	W916ZQ*	W91HY3**			3,149	2,243
					103,767 out of 132,746 filled requisitions	33,453 out of 78,814 filled requisitions

*Dominant ship-to address for USAMMC-SWA shipments.
**Dominant ship-to address for USAMMCE shipments.
NOTES: FY08 TAMMIS, GATES, RFID data. MRO type B only. Twenty Level III customers in Iraq, Afghanistan, and Kuwait. Must have arrival date (GATES APOD in-check or RFID ping at destination).
RAND MG929-D.1

FY2009 Consolidation and Transportation Times to Critical Locations

A comparison of FY2009 pick to APOE arrival times to those locations where the database records at least ten deliveries between October and December 2008 (see Figure E.1) demonstrates the performance advantage that USAMMCE and USAMMC-SWA have over DDKS, particularly where the Class VIII tender is utilized. It also shows that times from USAMMC-SWA to the USCENTCOM AOR are shorter than from USAMMCE but by less than one day to those areas that received high air freight volume, such as Balad and Baghdad.

Figure E.1
Comparison of Pick to APOE Arrival Times to Locations with at Least Ten Deliveries (October to December 2008)

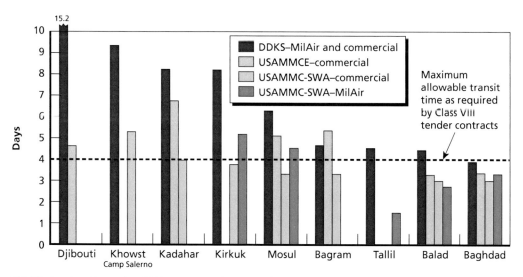

NOTES: October–December 2008 transportation times from origin to APOD where there were a minimum of 10 observations. USAMMCE and USAMMC-SWA commercial air data source: USTRANSCOM. USAMMC-SWA MilAir data source: USAMMC-SWA production reports. DDKS: Strategic Distribution database.
RAND *MG929-E.1*

Distribution Data Populations by Segment: USAMMCE and USAMMC-SWA

The sources of attrition in the data population for each distribution segment are given in Table F.1 (see next page). For USAMMCE, the main source of attrition in the data was whether an order hit RFID with a resulting valid destination arrival "ping" in the RFID system. The main source of attrition for USAMMC-SWA was whether the TCN matched to GATES and, again, if there was a valid arrival date stamp in GATES.

For USAMMCE FY2008, there were a total of 78,814 shipments to CENTCOM, of which 37,913 hit RFID and 28,765 had a valid destination "ping." For USAMMC-SWA FY2008, there were a total of 132,746 shipments to CENTCOM, of which 103,767 were in GATES and had a valid arrival date stamp.

Table F.1
Distribution Data Populations by Segment: USAMMCE and USAMMC-SWA

USAMMCE FY2008 Distribution Data Populations by Segment						
Variable	Label	N	Mean	50th Percentile	75th Percentile	95th Percentile
doc2rec	Document to depot receive	78,814	3.3	2	4	8
rec2iss	Depot receive to MRO cut	78,814	3.4	0	1	21
iss2pick	MRO cut to pick date	78,589	1.2	0	0	8
iss2wrdt	From MRO pick to RFID tag burn (triwall consolidated)	37,913	9.8	4	6	11
pic2wrdt	Pick date to RFID tag burn	37,762	8.6	4	6	11
transit	RFID tag burn to arrival at customer	27,058	4.7	4	5	14
total1	Document date to destination	28,765	18.9	12	19	51

USAMMC-SWA FY2008 Distribution Data Populations by Segment						
Variable	Label	N	Mean	50th Percentile	75th Percentile	95th Percentile
doc2rec	Document to depot receive	132,746	1.5	1	2	5
rec2iss	Depot receive to MRO cut	132,746	0.9	1	1	3
iss2pick	MRO cut to pick date	132,713	0	0	0	0
pic2poe	Pick date to arrive APOE	109,506	1.7	1	2	4
transit	Arrive APOE to arrive APOD	97,696	2.9	2	4	9
total1	Document date to destination	103,767	7.1	6	9	16

Origin-Destinations in the Data Analyzed for Transportation Performance

Figure G.1 shows the union of the origin-destination locations used in the transportation performance analysis. The first column on the left shows the USCENTCOM locations served by Class VIII commercial tender from USAMMCE. The second column from the left shows the USCENTCOM locations served by the Class VIII tender from USAMMC-SWA. The third column from the left shows those locations served by MilAir to USCENTCOM from USAMMC-SWA. The fourth column from the left shows those USCENTCOM locations served by MilAir and/or commercial tender from DDKS. The column on the far right shows those origin-destination pairs that are served by USAMMCE, USAMMC-SWA, and DDKS.

Figure G.1
Origin-Destinations in the Data Analyzed for Transportation Performance

USAMMCE via Class VIII commercial tender to:	USAMMC-SWA via Class VIII commercial tender to:	USAMMC-SWA via MilAir to:	DDKS via MilAir and commercial tender to:	Origin-destinations analyzed:
Iraq	**Iraq**	**Iraq**	**Iraq**	**Iraq**
Abu Ghraib	Al Asad	Al Asad	Al Sahra	**Baghdad**
Al Asad	Ali Al Salem Kuwait	Al Taqaddum	Al Taqaddum	**Balad**
Al Udeid	**Baghdad**	**Baghdad**	**Baghdad**	**Kirkuk**
Ali Al Salem Kuwait	**Balad**	**Balad**	**Balad**	**Mosul**
Baghdad	Bucca	Bucca	**Kirkuk**	**Tallil**
Balad	Habbaniyah	Camp Cropper	**Mosul**	**Afghanistan**
Bucca	**Kirkuk**	**Kirkuk**	**Tallil**	**Bagram**
El Kut	**Mosul**	**Mosul**	**Afghanistan**	**Kandahar**
Habbaniyah	Nasiriyah	**Tallil**	**Bagram**	**Khowst**
Kirkuk	**Tallil**	Tikrit	**Kandahar**	Djibouti
Mosul	Tikrit	**Kuwait**	**Khowst**	
Taji	**Kuwait**	Camp Arifjan	Jalalabad	
Tallil	Camp Arifjan	**Afghanistan**	Sharana	
Tikrit	**Afghanistan**	**Bagram**	Djibouti	
Kuwait	**Bagram**	Jalalabad		
Camp Arifjan	Camp Bastion	**Kandahar**		
Afghanistan	Jalalabad	**Khowst**		
Bagram	**Kandahar**	Sharana		
Kabul	**Khowst**	Djibouti		
Kandahar	Djibouti			
Bahrain				
Djibouti				
Bishkek, Kyrgystan				
Ashgabat, Turkmenistan				

RAND *MG929-G.1*

Bold blue text indicates location used for transportation comparison analysis.

Sensitivity Analysis: Cost per Issue over Time

During the course of the project, we were presented with background information on the establishment of USAMMC-SWA as a provisional organization to support the USCENTCOM AOR along with the associated discussion and planning for the ultimate closure of the depot. One metric that could be used to measure the cost-effectiveness of maintaining USAMMC-SWA is the cost per issue out of the depot. This appendix offers a detailed analysis of cost per issue as it would change over time at USAMMC-SWA within a changing USCENTCOM AOR.

As with the other sensitivity analyses developed in this project, it is important to recognize the size of operations in Afghanistan and Iraq. Figure H.1 shows the variation in the number of Class VIII issues to both Afghanistan and Iraq as related to the

Figure H.1
Number of Issues to Iraq at 1X of FY2009, Q1 Is Roughly Equal to Those to Afghanistan at 2X of FY1990, Q1

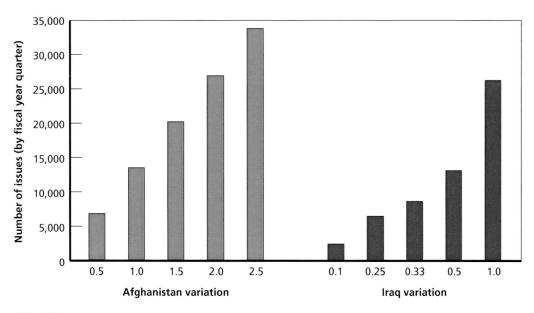

changing troop levels in each portion of the USCENTCOM AOR (as a fraction or multiple of "current"). Doubling issues in Afghanistan would be similar to the late 2008 level in Iraq.

Figure H.2 shows the variation in cost per issue from the period "current" through one possible future scenario made up of three additional periods that are each defined by changes in troop strength in Iraq and Afghanistan. The relationship between Class VIII requirements and troop strength is assumed to be linear and the changes in force size in each country are shown as a percentage of "current" for each country. As before, "current" is defined as the monthly average over October to December 2008. The first period is marked by Iraq at 0.33 and Afghanistan at 2.0. Period two is Iraq at 0.25 and Afghanistan at 2.5. Finally, period three is Iraq at 0.25 and Afghanistan at 0.5. The bars in the chart correspond to the primary vertical axis. The lower, blue portion of each bar is related to the fixed cost (or the minimum organic manning level) to operate USAMMC-SWA, while the red bar represents the variable costs associated with the manning provided through the EAGLE contract. The green (Iraq) and purple (Afghanistan) lines correspond to the secondary vertical axis on the right and display the number of issues to customers in each country. After period three, the cost per issue out of USAMMC-SWA almost triples. We suggest that USAMRMC develop cost per issue markers to inform decisions regarding whether or not to continue the USAMMC-SWA depot operation.

Figure H.2
As the Level of Effort in the CENTCOM AOR Decreases, It Becomes Increasingly Costly to Operate USAMMC-SWA

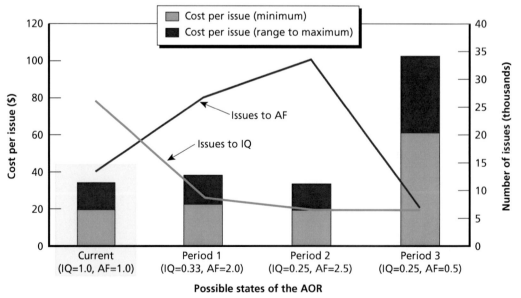

The operating cost for USAMMC-SWA has both variable and fixed portions. The fixed items are the cost of equipment and facilities as well as the cost of critical medical personnel that handle management and coordination functions. The variable item is the EAGLE contract for warehousing activities. Figure H.3 shows the variation in USAMMC-SWA operating cost that might be expected as the level of effort (number of troops deployed) to the USCENTCOM AOR changes. The minimum contract value for the EAGLE contract necessary to maintain full functionality of USAMMC-SWA is roughly $500,000 per year, which corresponds to a minimum total operating cost of USAMMC-SWA of just over $3 million per year.

Higher volume to both Iraq and Afghanistan drives down the cost per issue out of USAMMC-SWA, as shown in Figure H.4. As the USCENTCOM AOR continues to mature, the volume processed through USAMMC-SWA may drop to a level where maintaining an active, Class VIII depot in USCENTCOM is not financially efficient, nor fiscally responsible.

Figure H.3
Eagle Contract Represents the Variable Portion of the USAMMC-SWA Operating Costs; Minimum Contract Value Is ~$500,000

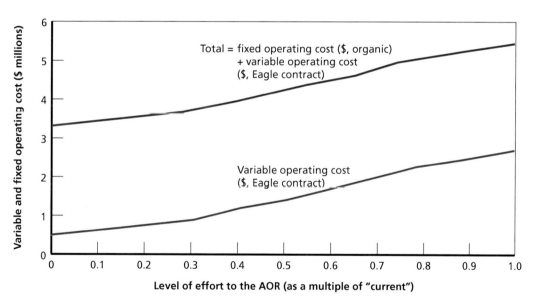

RAND MG929-H.3

Figure H.4
Range of the Cost per Issue Varies by Possible States of the AOR, but Is Lower as Iraq Increases and Afghanistan Increases

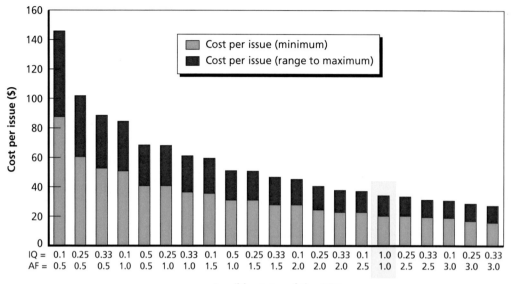

RAND *MG929-H.4*

Theater Express City Pair List

Figure I.1
Theater Express City Pair List (2009)

HDAM	1	Djibouti	OJHF	28	Prince Hasan	ORKK	55	Kirkuk
HECA	2	Cairo E.	OKAJ	29	Al Jaber	ORMM	56	Basrah Intl
HECW	3	Cairo W.	OKAS	30	Ali Al Salem	ORQW	57	Qayyarah West
LTAC	4	Esenboga	OKBK	31	Kuwait Intl	ORSH	58	Al Sahra
LTAG	5	Incirlik	DDKS	32	DDKS-Kuwait	ORTF	59	Tall Afar
OACC	6	Chakhcharan	OMAA	33	Abu Dhabi	ORTL	60	Tallil
OAFR	7	Farah	OMAM	34	Al Dhafra	ORUB	61	Al Kut
OAFZ	8	FAIZABAD	OMDB	35	Dubai Intl	OTBH	62	Al Udeid
OAHR	9	Herat	OMDM	36	Minhad	OYAA	63	Aden
OAIX	10	Bagram	OMFJ	37	Fujairah Intl	OYSN	64	Sana'a
OAJL	11	JALALABAD	OOMA	38	Masirah	UAFM	65	Manas
OAKB	12	Kabul	OOMS	39	Seeb	UTAA	66	Ashgabat
OAKN	13	Kandahar	OOTH	40	Thumrait	UTDD	67	Shamsi
OAMN	14	Maimana	OPJA	41	Jacobabad	UTSA	68	Navoi
OAMS	15	Mazar I Sharif	OPLA	42	Lahore	UTSL	69	Karshi Khanabad
OASD	16	Shindad	OPPI	43	Pasni	UTTT	70	Tashkent
OASG	17	Sheberghan	OPPS	44	Pseshwar	HKNW	71	Nairobi
OASL	18	Salerno	OPQT	45	Quetta	HSSS	72	Khartoum
OATN	19	Tarin Kowt	OPRN	46	Islamabad (Chaklala)	HAAB	73	Bole Intl
OAUZ	20	Kunduz	OPSM	47	Bandari			
OASA	21	Sharana	ORAA	48	Al Asad			
OAZI	22	Camp Bastion	ORAT	49	Al Taqqadum			
OBBI	23	Bahrain Intl	ORB4	50	Bashur			
OBBS	24	Shaikh Isa	ORDD	51	Balad			
OJAC	25	Amman, Jordan	ORBI	52	Baghdad			
OJAI	26	Queen Alia	ORBM	53	Mosul			
OJAM	27	Marka, Jordan	ORER	54	Erbil			

RAND *MG929-I.1*

References

Addison, D. L., R. M. Cocrane, P. A. Costello, M. G. Johnson, and J. M. Kissane, *Medical Materiel Readiness Assessment (2007)*, Report DL601T5, McLean, Va.: Logistics Management Institute, 2008.

Brew, M., *AMEDD Management of Class VIII: United States Army*, 2003a.

Brew, M., "The Transformation of Medical Logistics Since Operation Desert Storm," *The Association of the United States Army Magazine*, June 2003b.

Brew, M., and S. R. Campbell, "Modular Medical Logistics Support at the JRTC," *Army Logistician*, Vol. 39, No. 6, 2007.

Brooks, J. R., "Streamlining the OEF/OIF Class VIII Supply Chain," Thesis, Maxwell AFB: Air Command and Staff College, 2006.

Cintron, Angel E., and Phillip E. Livermore, *Medical Logistics—Pillar of Health Care Delivery*, 1993.

Donahue, Richard, and Mary Martin, "Dedicated Medical Logistics Management: Cornerstone to World Class Healthcare," *U.S. Army Medical Department Journal*, July/August 1995, pp. 26–31 (OCLC) 32785416, ISSN 1524-0436, 1995.

Galuszka, D. H., "Medical Logistics in a New Theater of Operations: An Operation Iraqi Freedom Case Study," U.S. Army Command and General Staff College, Fort Leavenworth, Kan., 2006.

Haddad, Sam E., "USAMMC-SWA Concept of Support Briefing," Camp As Sayliyah, Qatar, October 2008.

Joint Service J4 and Service 4s, Trip Report: Joint Service J4 and Service 4's AOR Logistics Visit, July 13–22, 2008.

Kistler, Thomas E., "A Case for the Separate Medical Logistics System," *Medical Bulletin of the U.S. Army, Europe*, December 1985.

Peltz, Eric, Kenneth J. Girardini, Marc Robbins, and Patricia Boren, *Effectively Sustaining Forces Overseas While Minimizing Supply Chain Costs: Targeted Theater Inventory*, Santa Monica, Calif.: RAND Corporation, DB-524-A, 2008.
http://www.rand.org/pubs/documented_briefings/DB524/

Office of the Under Secretary of Defense (Comptroller), *Memorandum: FY2008 Department of Defense (DoD) Military Personnel Composite Standard Pay and Reimbursement Rates*, April 2007.

Tyler, D. P., "Joint Supply Chain (Medical) Diagnostic Final Report," 2006.

United States Army, *Employment of the Field and General Hospitals*, FM 8-10-15, March 1997.

United States Central Command J4, Aerial Port Letter of Instruction, 2008a.

United States Central Command J4, Intra-Theater Airlift Letter of Instruction, 2008b.

United States Department of Defense Contract with Agility PWC to operate DDKS, Contract: SP3100-05-C-0020, September 2005.

United States Department of Defense Contract with EAGLE to support operations at USAMMC-SWA, Contract: W52P1J-08-C-0016, 2008.

United States Transportation Command J8, Uniform Tender of Rates and/or Charges for Transportation Services for TCAQ051908-01 Medical Supplies Class VIII, Reserved for USAMMC Use, Carriers: DHL, DLC, NAC, and UPS, August 2008.

Ursone, R. L., "Determine If a Need Exists for a Separate Medical Supply System," MSPR #33, 1988.

Workman, Dale H. August, "The Case for Separate Medical Logistics Management," *Army Logistician,* July-August 1985.